MATHEMATICS MASTERCLASSES
FOR YOUNG PEOPLE

Mathematics Masterclasses
for Young People

MICHAEL SEWELL

OXFORD
UNIVERSITY PRESS

OXFORD
UNIVERSITY PRESS

Great Clarendon Street, Oxford, OX2 6DP,
United Kingdom

Oxford University Press is a department of the University of Oxford.
It furthers the University's objective of excellence in research, scholarship,
and education by publishing worldwide. Oxford is a registered trade mark of
Oxford University Press in the UK and in certain other countries

© Michael Sewell 2017

The moral rights of the author have been asserted

First Edition published in 2017

Impression: 1

Published in the United States of America by Oxford University Press
198 Madison Avenue, New York, NY 10016, United States of America

British Library Cataloguing in Publication Data
Data available

Library of Congress Control Number: 2016954969

ISBN 978–0–19–880121–4

Printed and bound by
CPI Group (UK) Ltd, Croydon, CR0 4YY

FOREWORD

By Professor Sir Christopher Zeeman F.R.S.

I welcome this book. My own experience in teaching children was to give the Royal Institution Christmas Lectures in 1978. This was the first time since they began 180 years ago that they were devoted to mathematics. It had always been thought to be too difficult. In fact the children loved it. They loved the proofs. The great advantage of mathematics compared with science is that you can give the whole thing. And the children love that. In science you can't do all the experiments and so the children have to take these on trust, and there is always an element of magic about it, having to believe things you're told rather than seeing them for yourself. I found that 13-year-olds could understand the proofs, and loved them.

Professor Sewell has had much experience of teaching 9- and 10-year-olds. In this book he gives over 80 examples of mathematics, with over 70 diagrams. The book can be used by a teacher giving a class, or can be read alone by a bright individual. There are many individuals who love mathematics and are streets ahead of their colleagues in their ability to understand the subject, and to whom this book would be a treasure. The first step in knowing a subject is to know what it is about. After reading this book an individual will indeed know what mathematics is, and will be able to recognise a mathematical argument. So it is an ideal introduction.

I also liked the examples of spin-up. I can remember being fascinated by this when young. In fact in the very first lecture I ever attended at Cambridge the lecturer demonstrated this, and I eagerly looked forward to his mathematical explanation, but to my frustration he never gave it. So I had to wait several years before my frustration was resolved.

CONTENTS

LIST OF FIGURES

1 Introduction

For ten years, starting in October 2001, I taught a weekly mathematics session for an hour to a group of about ten selected ten-year-old pupils at Bisham School near Maidenhead, in the U.K.. In each year there were up to twenty sessions.

The principle which I tried to follow in choosing the topics was to find something which was not on the current syllabus for that age group and which might not be on the standard school syllabus that they would encounter in their most immediate future. I was pleasantly surprised to find, in trying to follow those guidelines, that I was able to introduce a variety of topics which entailed very little repetition in those ten years, and which often have some originality. Some additional material devised more recently, in the same spirit and accessible at a similar level, is also included here. The contents are also suitable for older 'young people'.

I am grateful to Mr Jim Cooke, the then Headmaster at Bisham School, for encouraging and facilitating this initiative. I wrote my notes for each session into a connected account immediately afterwards. The purpose here is to assemble those notes into a coherent version of many of the topics, and to extend them in certain directions, so that other teachers might find it helpful to use the material for their own able pupils. Of course, I am well aware that there are other initiatives of this type in the public domain, but it has not been my purpose to make direct use of material from other available sources. I am also grateful to Jacqueline Fairbairn for her contributions to the graphics required in several illustrations.

The underlying stimulus for me has come from the Mathematics Masterclasses initiative emanating from The Royal Institution, and encouraged particularly by Professor Sir Christopher Zeeman, F.R.S., from 1981 onwards. This has become a vigorous national scheme for various local age groups. Zeeman encouraged me to organise a series of Mathematics Masterclasses in Berkshire, which I did annually at the Department of Mathematics and Statistics of the University of Reading from 1991 to 2000. They continued at Wellington College for a short time, and then subsequently at Kendrick School in Reading, and then Holyport College. These are for thirteen-year-olds. I invited twelve speakers every year, each to deliver a two-and-a-half hour Masterclass on a Saturday morning. There were two parallel sessions on each of six mornings in the Spring Term. About thirty-six secondary schools each sent two pupils every week. Some of their teachers came to help with two exercise sessions which were interspersed with three sessions of expositions. The latter were delivered, over the ten years, by thirty-six speakers. In total, therefore, I arranged 114 such Masterclasses during that period.

Twelve of these Masterclasses were published in 1997 by Oxford University Press in a book which I edited. It is called *Mathematics Masterclasses: Stretching the Imagination*.

More recently, Masterclasses for pupils as young as eight have been arranged by Jim Cooke, Ron Lewin, myself and others at venues within the Royal Borough of Windsor and Maidenhead.

It would be reasonable to ask how the best use could be made of the material in this book by pupils, teachers, parents and other interested readers. I will indicate here some suggestions in this direction. The majority of the material has been constructed over a long period, so that it could be conveyed in class on an interactive basis, by question and answer sessions. Formal 'homeworks' were not required, and that is why no sets of formal 'exercises' are included here. This is not intended to be a 'textbook' in a conventional sense.

Instead, the emphasis throughout is to provide novel presentations of every topic. Important and basic mathematical ideas are presented in a way which often invites the reader to think 'out of the box'. That is often how new mathematics develops and grows. The book can be read by a single reader, with paper and pencil to hand for regular verification of the text, or by readers in discussion groups, who may hopefully have access to an adult if some assistance is required.

Care is taken in the first few sections to introduce some basic mathematical ideas, but quite quickly to also convey important mathematical results (e.g. Euler's formula) or currently unsolved problems (e.g. Goldbach's guess). This is to make it clear that more is waiting to be discovered, even about problems that can be simply stated. Encouragement is provided to use real situations (e.g. the 'foggy day problem') to show how ideas develop.

The underlying objective is, therefore, to provide examples which show the need for, and offer the opportunity for, original thinking. That is how real progress can be made in any endeavour. The importance of the idea of 'proof' in mathematics is deliberately emphasised.

Some important geometry about angles leads on to 'series', which describe an expanding school, and then to various practical situations which encourage independent observation, such as isosceles tiling, graphs which describe money changes, and novel Fibonacci series, which go not only forwards (increasing) but also (unconventionally) backwards through negative numbers. The purpose here, again, is to illustrate and encourage novel thinking.

After some further geometry applied, for example, to a rugby field and to 'lunes', the reader is invited to provide solutions to a coffee shop problem, and to an octet and then a mosaic of equal circles.

The objective of the book is to thus provide the reader with an understanding, via a diversity of frequently novel situations, of how mathematics really works and grows in practice. Hopefully, that understanding will encourage him or her to provide additional new examples. The term 'Masterclasses' is now a widely understood description for the type of material in this book.

2 Spin-Up

An interesting topic to offer at the start of a course is the *spin-up* phenomenon associated with different forms of spinning top. The dynamics of a top has been well studied and it has a long history, with textbooks dating back to the nineteenth century. The mathematics of it is not accessible to us at this level, but the experiment certainly is.

For our purpose there is a ready illustration available in modern supermarkets. Avocado pears enclose a seed or 'stone' which often (not always) has the desired axisymmetric shape, after cleaning. It is blunt at one end, but it has a relatively sharp point at the other end. In static equilibrium it will lay on its side on a horizontal table. If spin is then imparted to it by flicking the sharp and blunt ends in opposite directions with the fingers, it will stand up and spin on the point for a good fraction of a minute, or more.

Figure 1: Spin-up of acorn

I have exhibited this in the classroom and encouraged the children to buy their own avocado to try it. The same spin-up phenomenon can also be achieved with eggs, provided they are hard-boiled first. It does not work with a raw egg because the momentum given to the egg-shell by the fingers does not transfer quickly enough to the fluid interior. However, a hard-boiled egg is solid and axisymmetric about its long axis. Even though it does not have a point at one end like an avocado stone, it does have an asymmetry about every shorter (transverse) axis. This asymmetry offers a preferred 'sharper' end to support the spin.

An acorn often has a shape not dissimilar to the avocado stone, with a blunt end where it has been attached to its 'cup', and a pointed opposite end onto which it will spin-up after being started on its side. This experiment needs to be tried before the interior has detached from the shell by drying out. The photograph, taken by Simon Johnson of the Reading University Photographic Service, shows an acorn spinning on its point.

Smarties® are a UK brand of children's sweets in the shape of a rather flat ellipsoid, in which every cross-section perpendicular to the *short* axis has a circular shape (unlike the avocado stone or egg or acorn, in which every cross-section perpendicular to the *long* axis is circular), and they also exhibit the spin-up effect. When the spin is begun about the short axis, with the *Smartie*® flat on the table, it will spin-up and spin on the edge. In 1985 I demonstrated the spin-up of *Smarties*® in front of a five-man television crew in my office at the Department of Mathematics and Statistics of the University of Reading. The item on Meridian News at 6 p.m. that evening was prefaced by the interviewer remarking that 'A smartie-pants boffin at Reading University . . . '.

The avocado stone, egg and acorn all have an axis of symmetry which is longer than any diameter transverse to it. By contrast, the *Smartie*® has an axis of symmetry which is shorter than every transverse diameter.

A rugby ball is a version of the first three, but which is *also* symmetric about every transverse axis, and is an example of what is called a *prolate spheroid*. By contrast, the Earth's surface has a polar axis which is shorter than every equatorial diameter, and it is an example of what is called an *oblate spheroid*, like the *Smartie®*.

3 Subject Definitions

An early discussion is desirable about what should be the *names* of the more familiar parts of mathematics. The children suggested *maths, numeracy, mental arithmetic* and *mental maths*, perhaps in part reflecting what they had been told. I suggested the following:

> *Arithmetic* is the study of numbers and the relations between them, obtained by combinations such as adding, subtracting, multiplying and dividing.

> *Algebra* is a way of talking about numbers when we don't say in advance what particular numbers we are dealing with. We don't say, either because we don't know, or because we don't want to say so that statements can apply to a lot of different numbers. Algebra is thought to have been invented in about 1100 A.D. by an Arab called Al Habri. If this name is said quickly it can sound like 'algebra'.

> *Geometry* is the study of points, lines and curves in two dimensions, and surfaces as well when we are in three dimensions. The name means, taken literally, 'Earth measuring'. Related names are 'geography' (earth description) and 'geology' (the study of rocks).

> *Trigonometry* means, taken very literally, 'triangle measuring'. This is of particular interest to surveyors whose job is to make maps. One famous surveyor in the nineteenth century was Sir George Everest, in India.

4 Odds and Evens

A single suite of playing cards numbered 1 to 10, with the picture cards removed, can be used to discuss combinations as follows:

The ten cards are shuffled, two cards are selected blind, and we then multiply, add and subtract the pair of numbers selected. The question is then asked, 'in each of the three cases: is the result odd or even and does one of these answers predominate over several trials?'.

Sample multiplication results are:

product	9	9	12	12	36	27	70	16	2	20
even or odd	O	O	E	E	E	O	E	E	E	E

Sample addition results are:

sum	16	13	18	15	11	5	8	8
even or odd	E	O	E	O	O	O	E	E

Eventually, it emerges that

$$E \times E = E, \; E \times O = E, \; O \times E = E, \; O \times O = O,$$
$$E + E = E, \; E + O = O, \; O + O = E,$$
$$E - E = E, \; E - O = O, \; O - O = E.$$

From the first of these three groups we can see that there are three times as many even products as odd products, and from the third of these groups, we see that some negative numbers can be even as well as some positive numbers.

The playing cards having served their purpose to introduce the topic, and to lead to the results just above, we can now *prove* them by algebra as follows. Let m and n denote any integers, so that $2m$ and $2n + 1$ are typical even and odd integers. Then the proofs of properties of their typical products, sums and differences are as follows:

$(2m)(2n) = 2(2mn)$ and $(2m)(2n + 1) = 2m(2n + 1)$ are even;

$(2m + 1)(2n + 1) = 2(2mn + m + n) + 1$ is odd;

$2m + 2n = 2(m + n)$ and $2m + 1 + 2n + 1 = 2(m + n + 1)$ are even;

$2m + (2n + 1) = 2(m + n) + 1$ is odd;

$2m - 2n = 2(m - n)$ and $2(m + 1) - 2(n + 1) = 2(m - n)$ are even;

$(2m + 1) - 2n = 2(m - n) + 1$ and $2m - (2n + 1) = 2(m - n) - 1$ are odd.

The ability to provide a proof is crucial in mathematics.

5 Solving Equations

The equals sign, =, appears once in every equation, and the purpose of it is to express a *balance*, which the equation represents.

An equation may contain an *unknown* quantity and to find the value of that we may have to isolate it on one side of the equation.

To do this we usually need to perform manipulations on the equation. This process is like *washing one's hands*, in the sense that intrinsically we must always do the *same* thing to *both* sides of the equation, to sustain the balance. One washes both hands at the same time, not one and then, perhaps, the other.

Very simple examples are $n + 1 = 0, 3n = 12$ and $n/5 = 2$. To solve these, from each side we subtract 1, divide by 3 and multiply by 5, respectively. This gives $n = -1, 4$ and 10 in the respective cases, after performing those operations.

The same principle applies to every other equation, no matter how complicated it may be. In some cases several such operations will be required to isolate the unknown quantity.

6 Weighing the Baby by Algebra

I can stand on my bathroom scales, but my new grandson Daniel cannot sit on them without falling off. How can I use these scales, and some algebra, to weigh Daniel?

I need to introduce some labels for the weights. I choose m for me (or Michael, or man) and b for baby. The first measurement shows that $m = 13$ stones.

Next, I stand on the scales holding Daniel in my arms. This second measurement shows that $m + b = 14$ stones.

Now, I subtract m from both sides of the equation, which gives $m + b - m = 14 - m$, so that $b = 14 - m = 14 - 13 = 1$ stone. Thus, Daniel weighs 1 stone.

7 Prime Numbers

Before we define a *prime number*, it is informative to notice several other adaptations of this adjective, as follows:

(a) The 'prime of life' means the 'best time', as echoed in Sir Isaac Newton's quoted remark in old age that, between the ages of 20 and 25 'I was in the prime of my age for invention' (of mathematical theories).

(b) The 'Prime Minister' is 'chief' Minister in the sense of being the most important.

(c) A 'prime' cut of beef is one of the best flavoured cuts.

(d) The 'primary colours' are those members of the spectrum which cannot be made from other colours.

(e) A 'premier' league of footballers contains the best teams.

(f) This example is a verb, not an adjective. To 'prime' a surface for painting means to prepare it, by giving it a coat of 'primer'. To 'prime' a gun means to prepare it for firing. To 'prime' an engine means to get it ready for use.

Turning now to numbers, we first agree not to designate 1 as either prime or not prime, because this would not be helpful. 1 is too bland to merit either adjective. For example, it does not *do* anything in multiplication or division.

We now *define* a *prime number* to be an integer (whole number), which is divisible only by itself, so that it has no other factors.

Immediately we can say that *no even* numbers are prime, because each has 2 as one of its factors, by definition.

Only *some* odd numbers are prime, such as 7 and 29, but there are other odd numbers which are not prime, such as $21 = 3 \times 7$ and $65 = 5 \times 13$.

The primes between 1 and 50 are

$$2, 3, 5, 7, 11, 13, 17, 19, 23, 29, 31, 37, 41, 43, 47.$$

Number theory is a large subject, and there are some intriguing properties of prime numbers which are part of it.

For example, there are some pairs of *adjacent* numbers, which are products of sets of prime numbers, and the sum of each of the two sets is the *same*.

A simple example is the pair

$$5 \text{ and } 6 = 2 \times 3, \text{ for which } 5 = 2 + 3.$$

A less simple example is the pair

$$714 = 2 \times 3 \times 7 \times 17 \text{ and } 715 = 5 \times 11 \times 13,$$

for which $2 + 3 + 7 + 17 = 29 = 5 + 11 + 13$.

We notice that each of these two examples uses a *consecutive* set of primes in all. The first set is just 2, 3, 5. The second set is

$$2, 3, 5, 7, 11, 13, 17.$$

It has been calculated that there are 26 pairs of consecutive numbers below 20 000 which have the property that the sum of their prime factors is the same.

A different facet of prime numbers is revealed by the facts that $5 \times 6 = 2 \times 3 \times 5$ is the product of the first three prime numbers, and

$$714 \times 715 = 2 \times 3 \times 5 \times 7 \times 11 \times 13 \times 17$$

is the product of the first seven prime numbers.

8 Don't Jump to Conclusions

Here we ask the question: does the formula $N = n^2 + n + 17$ deliver a *prime* number N for every positive integer n?

As we have seen, an integer is a whole number (not a fraction or decimal) and a prime number is an integer which is divisible (i.e. leaving no remainder) by itself only. Put otherwise, we may say that a prime number has no factors.

If we evaluate $N = n^2 + n + 17$ long-hand, we find that the following values of n,

$$1\ 2\ 3\ 4\ 5\ 6\ 7\ 8\ 9\ 10\ 11\ 12\ 13\ 14\ 15\ 16\ 17,$$

generate the following corresponding values of N:

$$19\ 23\ 29\ 37\ 47\ 59\ 73\ 89\ 107\ 127\ 149\ 173\ 199\ 227\ 257\ 289\ 323.$$

Examining, in turn, the numbers in this list shows that each of the first 15 N have no factors. In other words, N is prime for $n \leq 15$. That rather large such number of those results might tempt us to conclude that the starting formula delivers a prime number N for every n.

But this is not so because closer examination shows us that $289 = 17^2$ and $323 = 17 \times 19$. So neither 289 nor 323 is a prime number, because they each have a pair of factors which are integers.

This is a reminder that every assertion in mathematics, even if it looks plausible, requires a proof.

9 Euler's Formula

Leonard Euler was a Swiss mathematician who lived from 1707 to 1783. He was one of the most famous and prolific mathematicians of all time. He worked in St Petersburg (1727–1741 and 1766–1783) and Berlin (1741–1766).

Among many other things, Euler pointed out that the formula

$$E = n^2 + n + 41$$

delivers a prime number E for every integer $n = 1, 2, 3, \ldots 37, 38, 39$; but for $n = 40$ we find that
$E = 40^2 + 40 + 41 = 1681 = 41^2$, so this is *not* prime and is, in fact, a square number.

So it would have been false to jump to the conclusion that $n^2 + n + 41$ always delivers a prime number for every integer n, even though it does so for the first 39 of them.

10 Goldbach's Guess

This is a famous conjecture, which has *never been proved*. It was stated by Christian Goldbach in a dialogue with Euler in 1742. There are several forms of it, but a simple one is the following:

'Every even number greater than 2 is the sum of two primes.'

Examples are $8 = 5 + 3$, $10 = 5 + 5 = 7 + 3$, $12 = 7 + 5$, $14 = 7 + 7 = 3 + 11$, $20 = 13 + 7 = 17 + 3$, $42 = 23 + 19 = 29 + 13 = 31 + 11 = 37 + 5$.

11 Perfect Numbers

Pythagoras of Samos (c. 570 B.C.–c. 495 B.C.) is supposed to have introduced this idea. A *perfect number* is defined to be a number which is the sum of its divisors. The first two are

$$6 = 1 \times 2 \times 3 = 1 + 2 + 3 \text{ and}$$
$$28 = 1 \times 2 \times 14 = 1 \times 4 \times 7 = 1 + 2 + 4 + 7 + 14.$$

The ancient Greeks knew about these two, and two more, namely 496 and 8128.

Seventeen centuries later a fifth was discovered, namely 33550336, which has 8 digits. Currently 37 are known, the largest having 1 819 050 digits.

All 37 are even. Nobody knows whether there is an odd one.

12 Euclid's Theorem

Euclid was a Greek mathematician working in Alexandria (now part of Egypt) around 300 B.C. His famous writings have survived in thirteen books, now called Euclid's *Elements*. The following result is cited to be in *Elements* IX 20, by G.H. Hardy in his famous small book called *A Mathematician's Apology*. This was first published in 1940 by Cambridge University Press, and reprinted several times subsequently.

Euclid's Theorem states that there is no largest prime number.

The proof is as follows. It uses a classical method called *reductio ad absurdum*, which means 'reduce to an absurdity', and proceeds by showing that the opposite of the desired result implies a nonsense.

Suppose, then, that there *is* a largest prime number, and denote it by P. Of course we do not know its value at this stage. We use every prime number up to and including P to construct another number, which we call Q, by the following definition:

$$Q = (2 \times 3 \times 5 \times 7 \times 11 \times \ldots \times P) + 1.$$

Each of the prime integers from 2 up to P appears once in this definition, and none of them divides into Q because there is always a remainder 1 when we try to do that division. So if Q is not prime, it must be divisible by a prime which is larger than P.

But if Q is prime, it is obviously larger than P.

These two conclusions together contradict our starting assumption that P is the largest prime.

So that assumption, that there is a largest prime, is wrong.

So there is no largest prime, and there must be an infinite number of prime numbers. They just go on and on.

Euclid's Theorem is simple to state, but even so it requires proof. The *method* of proof, above, by *reductio ad absurdum* is profound and powerful.

13 Mathematical Symbols

We have already used some of these, including the equals sign =, the addition sign +, the minus sign − and the multiplication sign ×.

Before going further, it is prudent to notice some other common signs. The symbol < means strictly less than, as in $7 < 12$; and > means strictly greater than, as in $66 > 58$.

Associated with these are ≤ and ≥. The first of these means 'less than or equal to'. For example, if y denotes the number of days in a year, then $y \leq 366$ because we must allow for leap years. The symbol ≥ means 'greater than or equal to', which can be used to express the fact that the number f of the days in February is such that $29 \geq f \geq 28$. In other words, as the well-known mnemonic tells us, 'February has 28 days clear, and 29 in each leap year'.

The symbols ≪ and ≫ mean, respectively, 'very much less than' and 'very much greater than'. If c is the age of a child and p is the age of an old-age pensioner, then $c \ll p$ and $p \gg c$.

The symbol ≠ means 'not equal to', so that we can write 1 metre ≠ 1 yard. This use of a slanting line across a symbol can be seen in traffic signs which indicate that something should *not* be done. The symbol ≈ means 'approximately equal to'.

And 'infinity' is written as ∞. In fact ∞ is not a number at all, but it is an example of what is called a 'limit', to which (in this example) the fraction $1/n$ tends as the number n in the denominator approaches 0.

14 Medicine Problem

This is a short exercise in formulating a real-life problem mathematically and then solving it.

Each day I have to take 5.5 mg of medicine, in the form of pills which are available in two strengths: pill A contains 3 mg and pill B contains 1 mg (mg means milligrams).

I deal with this by choosing to take 6 mg one day and 5 mg the next day, and then by repeating this two-day cycle. How many pills of each type do I need in a 28-day supply from the pharmacy?

Every *two* days I need 11 mg. This can be arranged as

$$11 = (2 \times 3) + [1 \times (3 + 1 + 1)] = 2A + (A + 2B) = 3A + 2B.$$

So every 28 days I need

$$14 \times (3A + 2B) = 42A + 28B.$$

Therefore I need 42 of A = 3 and 28 of B = 1.

To check that my strategy delivers the correct outcome of

$$28 \times 5.5 = 140 + 14 = 154 \, \text{mg},$$

I observe that

$$42A + 28B = (42 \times 3) + (28 \times 1) = 154.$$

So the strategy works.

15 Dramatic Dates

This section was written on 1 January 2011, which can also be written as 1-1-11. There were other notable dates during that year, such as 11-1-11, 1-11-11 and, most symmetric of all, 11-11-11.

When time is introduced as well, the time at 2 minutes and 3 seconds past 1 o'clock in the morning of 4 May 2006 could have been written as 01.02.03 04-05-06. And the time at 6 minutes and 6 seconds past 6 o'clock in the morning of 6 June 2006 could have been written as 06.06.06 06-06-06.

In a lesson on 20 November 2002 it was noticed that the date could be written as 20-11-02. Numbers which have this symmetrical feature of being the same whether they are read forwards or backwards are called 'palindromic'. My Chamber's 1952 Dictionary gives, as an example of a 'palindrome', Adam's reputed first words to Eve: 'Madam, I'm Adam'.

Earlier palindromic dates in 2002 were 20-1-02 and 20-2-02, which could also be written as 20-02-2002.

More elaborate palindromes can be constructed by adding a particular time to certain dates. I saw a photograph in a newspaper of a baby called Lily Fry who had been born at 2 minutes past 8 o'clock on the evening of 20 February 2002. On her wrist was visible an identification tag which read 20.02 20-02-2002. Perhaps she should have been christened Lily Palindromeda Fry.

The sentence: 'Was it a cat I saw?' is another palindrome.

An alternative adjective to 'palindromic' could be 'reflective'.

16 The Foggy Day Problem

I was driving to Cambridge on a foggy day and I found that it was possible to see the Sun's disc, and to look directly at it without danger to my eyesight because of the fog. The Sun seemed to be about the same size as a full moon, but we know that this is not really so.

The illusion arises because the Sun is much further away than the Moon. We can introduce the mathematical idea of *similar triangles*, with some data about the Moon from an atlas, to determine the size of the Sun.

We draw a long, thin isosceles triangle ('isoskeles' is Greek for 'equal legs'), and mark along the long sides m and s for the distances, of the Moon and Sun respectively, from my eye, which is at the vertex. We also mark the diameters M of the Moon, and at the base of the triangle, S of the Sun. The diagram shows a schematic version of this juxtaposition (a *scale* drawing would have to go off the page, or make the Moon so small that it would be scarcely visible, because the actual distances of Sun and Moon are so different, as are their relative sizes).

The fact that we now have two similar triangles with the same angle at the vertex expresses the property that the ratio diameter/distance is the same for both Sun and Moon. That is,

$$\frac{S}{s} = \frac{M}{m} \text{ or } \frac{S}{M} = \frac{s}{m}, \text{ which implies } S = \frac{sM}{m}.$$

The atlas tells us that $M = 2172$ miles and $m = 239\,000$ miles.

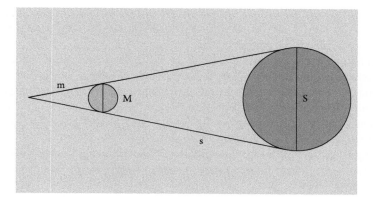

Figure 2: Sun–Moon triangle

The value of m is actually slightly variable because the Moon's orbit around the Earth is an ellipse and not a circle, but this variation of m is small and we ignore it in our approximation.

We also know that $s = 93\,000\,000$ miles, approximately. So,

$$S = \frac{sM}{m} = \frac{93000000 \times 2172}{239000},$$

which is approximately $\frac{93000 \times 2000}{200} = 930\,000$ miles without much error. So the diameter of the Sun is only a little less than one million miles and the ratio of Sun/Moon diameters and distances is

$$\frac{S}{M} = \frac{s}{m} = \frac{93000}{239},$$

which is approximately $\frac{100000}{250} = 400$. Thus, our Sun is 400 times bigger than the Moon and 400 times further away.

There was a lunar eclipse at 7.45 a.m. on 21 December 2010 as the Moon moved into the Earth's shadow made by the Sun. This was the first time since 1638 that a lunar eclipse had occurred on the day of the Winter Solstice.

17 Angles

What *is* an angle? We mentioned one in *The Foggy Day Problem*, but we ought to have a definition of *angle*.

First we have to decide whether we wish to work in two dimensions, for example, on a flat page or a curved surface, or in the three dimensions of our living space.

In *three* dimensions we might need to talk about the angle which would be traced out by a movable straight rod pivoted at a fixed point, and moved around so that it returns to its starting position. The space which could be traced around in this way is called a *solid angle*.

For simplicity we shall work in two dimensions on a plane, so that we can draw all diagrams on a flat page. Then we can make the following *definition*. An *angle* is the space between two directions out from a point.

As soon as we begin to draw the diagram we realise that every pair of directions from a point actually creates *two* angles, and that usually one of these is smaller than the other (unless they are both equal because the two directions are directly opposite along the same straight line). The diagrams show all possible cases.

The smaller angle is called either *acute* or *obtuse*. Acute is smaller than obtuse, unless they happen to be equal, when each is called a *right angle*.

The angle which completes the circle in each of these three cases is called a *reflex* angle.

In the special case when the obtuse angle is equal to the reflex angle, both of these are the *angle on a straight line*.

We can sense an immediate need to *measure* angles, so we need to define a *unit* of measurement. There are several different available options in use for such units and the choice of option depends on the purpose.

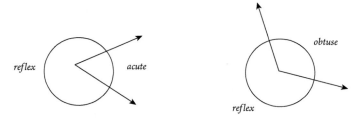

Figure 3: Acute and obtuse angles, with their reflex complements

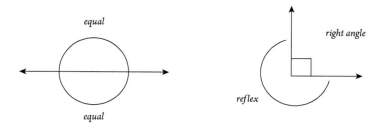

Figure 4: Two equal angles on a straight line, and a right angle

A *clock face* illustrates an obvious purpose, so we use this as a starting point for the discussion. Historically, clock faces were commonplace long before digital clocks. A digital clock offers less explicit information than a clock face, because it does not show how the current time is related to all the other possible times.

A clock face is a full circle which is divided up into a number of equal angular intervals between different directions. Most commonly it represents a twelve-hour clock. This means that the circle is divided into twelve equal divisions. The angle between each pair of these divisions represents *one hour*. The short hand of the clock tracks these.

But hours are not convenient for shorter times, so the same clock face is also divided up into 60 equally spaced directions. The angle between each pair of these represents *one minute*. The long hand of the clock tracks these.

Further subdivisions of the clock face are needed for shorter times. For example, between each pair of minute directions there are 60 *second* directions. These are not commonly marked on the most familiar clock faces, but if they were, there would need to be $60 \times 60 = 3600$ markers.

Another common unit of angle measurement is called the *degree*. The definition is that a full circle is divided up into 360 equal degrees. In other words, *one full revolution* contains 360 degrees. Degrees are the simplest angle measurement used in geometry.

Because a single revolution contains 360 degrees in geometry (which means 'earth measuring') and 3600 seconds 'of arc' on the clock face (i.e. in 'time measuring'), there is a sense in which 1 degree of arc = 10 seconds of time.

There are some special angles which occupy less than a single full revolution. An *angle on a straight line* is occupied by 180 degrees, so that the *angle in a semicircle* is 180 degrees.

A so-called *right angle* contains 90 degrees, so it is halfway round to a straight line, or a quarter of the way round a single full revolution. On a diagram a right angle is often denoted by a small square box at that corner.

The symbol for *degrees* is a superscript circle, as in 180° for a half-revolution. As we shall see, the same symbol is used for degrees in temperature measurements, even though the meaning is quite different.

There is another basic way of describing the measurement of angles which is fundamentally different from the use of degrees. There is a sense in which it is arbitrary to choose 360 degrees as the measure of a single full revolution, in the way that is described above.

An alternative measure of angles stems from the intrinsic experimental fact that, for *any* circle, the ratio of the circumference C to the diameter D is always the same number:

$$\frac{C}{D} = 3.14159265358....$$

This is an *irrational* number, which means that it cannot be expressed as an exact fraction, although the ratio $\frac{22}{7}$ is often used as a good approximation. The Greek letter

$$\pi = 3.14159265358....$$

is always used as the symbol for this number, and if we introduce the radius R of the circle via the fact that $D = 2R$, we find that

$$C = 2\pi R$$

for any circle.

This equation is used to define a *new* measure of angle by saying that there are

$$\frac{C}{R} = 2\pi \ radians$$

in a single full revolution, so that

$$2\pi \ \text{radians} = 360 \ \text{degrees},$$

and therefore,

$$1 \ \text{radian} = \frac{180}{\pi} \approx \frac{180 \times 7}{22} \approx 57.27 \ \text{degrees}.$$

Thus, the choice of 2π to be the number of radians in a full circle is *not* an arbitrary one, unlike the number of degrees. But each of these two choices has its convenience value.

Historical values of π which are said to have been calculated include:

3.1418 by Archimedes in 250 B.C.,

3.14166 by Ptolemy in 150 A.D.,

3.14159 by Tsu Chung Chi in 480 A.D.,

3.141818 by Fibonacci in 1220 A.D.,

3.141592653877932 by Newton in 1665 A.D.,

440 decimal places by Rutherford in 1853 A.D.,

620 decimal places by Ferguson in 1946 A.D.

Methods of calculating π in the classroom include the use of a ruler and a piece of string applied to cups, saucers, plates and (cleaned) pizza platters, to measure C and D for each directly, and then calculating the ratio $\frac{C}{D}$.

18 Angles Inside a Triangle

After our introduction to *Angles*, we state and prove a serious mathematical result, using the classical Theorem/Proof style which we indicated in the proof of Euclid's Theorem in Section 12. We use degrees instead of radians in this discussion.

Theorem:
The sum of the interior angles in any triangle is 180°.

Proof:
Draw any triangle (having no special features) and label the interior angles a, b, c. Then the required equation is $a + b + c = 180°$. Construct the exterior angles A, B, C by extending the sides as shown, cyclically.

Now write down what we know, which is that $A + B + C = 360°$ because this is the *whole* way round a cycle; and also that $A + a = 180°$, $B + b = 180°$, $C + c = 180°$ because each of these is half-way round.

Adding these last three equations gives $A + B + C + a + b + c = 540°$. Subtracting $A + B + C = 360°$ from corresponding sides of this last equation leaves $a + b + c = 180°$.

This is the required result.

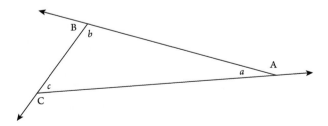

Figure 5: Angles in a triangle

19 Angles Inside a Quadrilateral

A polygon is a straight-sided closed diagram with several sides. For example, a triangle is a three-sided polygon.

A quadrilateral is a four-sided polygon; special examples of which are the square, rectangle and parallelogram.

Every interior angle of a square or rectangle contains 90°, so the sum of all four interior angles is 4 × 90 = 360°.

A parallelogram, having interior angles a, b, c, d, has the property that opposite angles are equal, i.e. $a = c$ and $b = d$, and also $a + b = b + c = c + d = d + a = 180°$, because the sum of the angles on a straight line is 180°. Therefore,

$$a + b + c + d = (a + b) + (c + d) = 180° + 180° = 360°,$$

so the sum of the interior angles of a parallelogram is 360°. Does every quadrilateral in which all the sides may be unequal, have this property?

Before answering this question, we re-examine the definition of a quadrilateral. It is certainly a four-sided polygon, as stated above, but when we move from triangles to quadrilaterals we need to recognise another possibility.

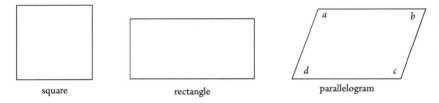

square   rectangle   parallelogram

Figure 6: Square, rectangle and parallelogram

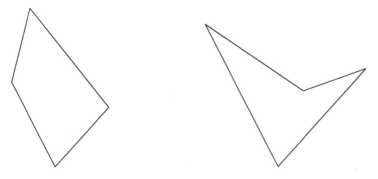

Figure 7: Convex and indented quadrilaterals

Our first instinct may be to suppose that every quadrilateral is *convex*, so that none of the corners are indented; but a second possibility is that one corner of a quadrilateral may be *indented*. Figure 7 shows these two cases.

Both of these shapes are quadrilaterals, because each has four sides, and the four sides can be of unequal lengths. It is self-evident that *at most* one corner of a quadrilateral can be indented.

In fact, there is a third possibility. The indented corner of the quadrilateral could cross one of the opposite two sides, thereby creating two intersections. The reader will be able to supply that diagram. The following theorem also applies to that case, as well as to the first two cases.

Theorem:
The sum of the interior angles in any convex or indented quadrilateral is 360°.

Proof:
For a convex quadrilateral, the proof follows the method of the previous proof about triangles. Label the interior angles a, b, c and d as shown.

Construct the *exterior* angles A, B, C, D by extending the sides in turn, cyclically as shown. In sequence they make one full revolution of $A + B + C + D = 360°$. We also have the property that the sum of the angles on a straight line is 180°, so that $A + a = B + b = C + c = D + d = 180°$. Therefore,

$$(180 - a) + (180 - b) + (180 - c) + (180 - d) = 360, \text{ and so,}$$
$$a + b + c + d = 360°.$$

This proves the result for the *convex* quadrilateral.

Turning now to the indented quadrilateral, we label the interior angles a, b, c, d as before, but now we have chosen $c > 180°$ as shown in Figure 9.

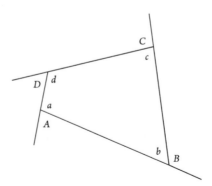

Figure 8: Angles of a convex quadrilateral

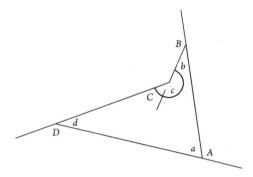

Figure 9: Angles of an indented quadrilateral

This means that $c - C = 180°$ at the indented corner, but $a + A = b + B = d + D = 180°$ at the other three corners. Hence,

$$a + b + c + d = 180 - A + 180 - B + 180 + C + 180 - D$$
$$= 720 - (A + B + D) + C.$$

But the full revolution property this time is $A + B - C + D = 360°$, so again, we deduce $a + b + c + d = 360°$.

Therefore, the sum of the interior angles of the indented quadrilateral is 360° also.

Q.E.D.

Q.E.D. means '*Quod erat demonstrandum,*' which is Latin for 'Which was to be proved'. It was customary in the older textbooks to write Q.E.D. at the end of proofs, but more recently a small square is commonly printed instead of Q.E.D.

20 Angles Inside a Polygon with n Sides

We have proved that the sum of the interior angles of a triangle is 180° and of a quadrilateral (both convex and indented) is 360°. These are the cases with $n = 3$ and 4. What is the sum of the interior angles when the polygon has any number n sides, whether it be convex or indented?

Theorem:
The sum of the interior angles of any convex or indented polygon with n sides is $2n - 4$ right angles.

Proof:
There are at least two different constructions which provide a proof of this result. The first one imitates the proofs given above for the cases $n = 3$ and $n = 4$. Suppose there are x convex corners and v concave corners. Denote the interior angles at the convex corners by $a_1, a_2, \ldots, a_x$ (all less than 180°), and at the concave corners by $b_1, b_2, \ldots, b_v$ (all greater

than 180°). Denote the associated exterior angles by $A_1, A_2, \ldots, A_x$ and $B_1, B_2, \ldots, B_v$. Then $a_i + A_i = 180°$ for each $i = 1, 2, \ldots, x$, and $b_j - B_j = 180°$ for each $j = 1, 2, \ldots, v$.

The sum of all the interior angles is, therefore

$$a_1 + \ldots + a_x + b_1 + \ldots + b_v = 180x - (A_1 + \ldots + A_x) + 180v + (B_1 + \ldots + B_v).$$

But the full revolution property is $A_1 + \ldots + A_x - (B_1 + \ldots + B_v) = 360$,

so the sum of all the interior angles is $180(x + v - 2) = 90(2n - 4)$ degrees, or $2n - 4$ right angles, because $x + v = n$.

Q.E.D.

This first proof illustrates, incidentally, the convenience of choosing suggestive notation in mathematics. We used x for convex and v for concave. To give illustrative diagrams we have to make particular choices, and we have chosen $n = 6$ with, for the indented hexagon, $x = 4$ and $v = 2$ (in other words, four convex corners and two concave corners as shown).

The second proof, for a convex polygon, begins with a construction which joins every vertex to an arbitrary internal point. We can always do this by the definition of convex polygon. Figure 10 shows an example of the construction for an irregular convex hexagon. If the polygon has n sides, the sum of the internal angles in all the n triangles so formed is $180n$ degrees, and this will include the 360 degrees at their common point inside the polygon. Therefore, the sum of the internal angles in the polygon itself is

$$180n - 360 \text{ degrees} = 90(2n - 4) \text{ degrees} = 2n - 4 \text{ right angles.}$$

If the polygon is not convex, consider the example of a pentagon with just one indented vertex. There is one internal reflex angle a_5, and four internal acute or obtuse angles a_1, a_2, a_3, a_4. Now complete a parallelogram by adding two more sides opposite a_5 so that the

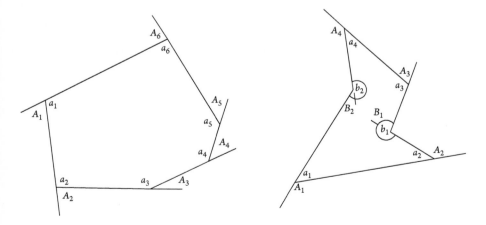

Figure 10: Angles of a convex hexagon and a non-convex hexagon

internal angles in this parallelogram are $360 - a_5$ (twice) and b_1 (twice), as shown. The two convex polygons so constructed have the properties

$$a_1 + a_2 + a_3 + a_4 + 2b_1 + 360 - a_5 = ((2 \times 5) - 4)90,$$
$$\text{and } 2b_1 + 2(360 - a_5) = ((2 \times 4) - 4)90,$$

which combine to give $a_1 + a_2 + a_3 + a_4 + a_5 = ((2 \times 5) - 4)90$.

There could be *two* indented vertices in a pentagon, rather than one as we have illustrated, and the same construction could be applied to both indents to create a new but convex pentagon, and thus facilitate the same proof.

An n-sided polygon might have $n - 3$ indented angles, and similar constructions may be applied for any n.

[Setting down four points 'usually' defines the corners of a quadrilateral, which might be convex (no indents) or concave (one indent). But setting down five points *always* includes four, which *do* make a *convex* quadrilateral.]

21 Method for Finding the Centre of a Circle

Finding the centre of a circle when we have only a saucer and a ruler, and no compasses.

(a) Draw round the saucer—this gives the circumference of the circle.

(b) Place the ruler across the circle, draw a line along both sides, measure to find the lengths of the chords so formed and then mark the midpoints M_1 and M_2 of these chords.

(c) Join these midpoints. This join M_1M_2 must go through the centre C of the circle, by symmetry, so we have drawn one diameter.

(d) Repeat (b) and (c) with the ruler in another position, not parallel to the first one, to find another pair of midpoints N_1 and N_2. This produces a second diameter N_1N_2.

(e) The intersection of these diameters will be the centre of the circle.

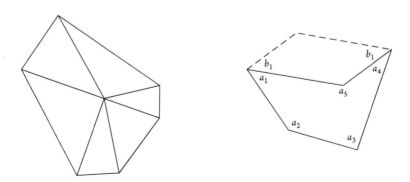

Figure 11: Convex hexagon and angles of a convexified pentagon

22 Angles in a Sector of a Circle

A *semi-circle* is half of the circumference of a circle. A *sector* of a circle is part of the circumference, which may be more or less than a semi-circle. A *chord* of a circle is a straight line joining any two points on the circumference. The *diameter* is that particular chord, through the centre, which divides the circle into two equal parts, i.e. into two semi-circles.

Now we imagine that we draw any chord, not necessarily the diameter, and thus define two sectors. We use that chord as the base of a triangle whose third corner is on the circumference in one or other of the two sectors.

The angle at a vertex in the smaller sector is greater than a right angle; at a vertex in the larger sector it is less than a right angle; and at a vertex in a semi-circle it is a right angle.

Many triangles can be drawn in any particular sector, and the angles at a vertex on the circumference are the same for every such triangle in that sector.

These all offer worthwhile exercises in the use of compasses for drawing the circles, rulers for drawing the triangles and protractors for measuring the angles at the circumference.

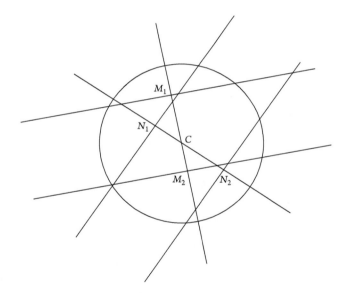

Figure 12: Centre of a circle by intersection of diameters

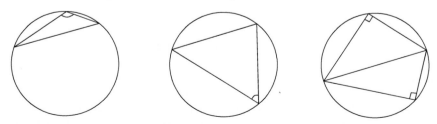

Figure 13: Obtuse, acute and right angles, in a sector of a circle

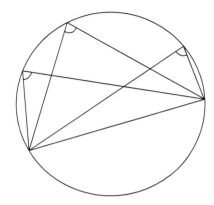

Figure 14: Equal angles in the same sector

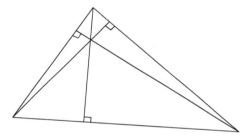

Figure 15: Orthocentre of a triangle is the intersection of the altitudes

23 Orthocentre

For any triangle, this is the point where perpendiculars 'dropped' from each of the three vertices, to the opposite side, intersect. This is a worthwhile ruler and compass construction.

First, from any vertex as centre, find the two intersections of a circular arc with the opposite side. Secondly, find the perpendicular bisectors of those two intersections. Thirdly, repeat the construction from another vertex. The intersection of those two perpendiculars defines the *orthocentre*. The third perpendicular, constructed in the same way as the first two, will pass through it.

24 Arithmetic Progression

An *integer* is a whole number, with no fractions. The sum of the first three integers is easily found $(1 + 2 + 3 = 6)$, and the sum of the first nine integers takes a little more time. It is $1 + 2 + 3 + 4 + 5 + 6 + 7 + 8 + 9 = 45$. The sum of the first 99 integers would take *much* more time if we went straight at it, and this problem provides a good example of how mathematicians often prefer what has been called 'The Higher Laziness', which means 'constant hard work in search of the easy way'. It is frequently fruitful.

To illustrate this we first give the answer a label, by writing
$1 + 2 + 3 + \ldots \ldots + 97 + 98 + 99 = A$, so that A stands for that sum. Then
$99 + 98 + 97 + \ldots \ldots + 3 + 2 + 1 = A$ is the same thing written backwards.

We now add both expressions vertically which gives

$$100 + 100 + 100 + \ldots \ldots + 100 + 100 + 100 = 2A.$$

In this expression there are 99 terms on the left, so that $2A = 9900$, and $A = 4950$.

The same technique can be employed if the last integer in the sum is any integer, say N, instead of 99, and we find $2A = (N + 1)N$, so that $A = \frac{1}{2}N(N + 1)$ is the sum of the first N integers.

The type of sequence just described, of a set of regularly spaced numbers added together, is called an 'arithmetical progression'. A diagrammatic version of the procedure is illustrated here for the sum of the first three integers. The dots just represent blanks.

1	1	1	.	.	.	1
1	1	.	.	.	1	1
1	.	.	.	1	1	1

This gives $2A = (3 + 2 + 1) + (1 + 2 + 3) = 6 \times 2$, so that $A = 6$.

Figure 16: Old School, King's School Grantham

A more general version can have *any* starting number, say *a*, and *any* difference *d* between successive numbers, so that

$$A = a + (a + d) + (a + 2d) + \ldots + (a + (n - 1)d) + (a + nd).$$

Reversing the order of the terms in this, we have

$$A = (a + nd) + (a + (n - 1)d) + (a + (n - 2)d) + \ldots + (a + d) + a.$$

There are $n + 1$ terms in each of these series, and the structure is such that if we add vertical pairs of terms we get the simple result

$$A = \frac{1}{2}(2a + nd)(n + 1).$$

25 Leavers from an Expanding School

A graph which shows how the number of pupils in a school depends on time will normally be what is called a 'step function'. There will be a vertical step each September, indicating the difference between the number of new entrants and the number of recent leavers, and horizontal lines showing the level of the population between adjacent Septembers. For an expanding population the steps will be upwards. Over a period of years the size of the steps may change. Over a long period of years it may be mathematically convenient, as an approximation, to replace the step function (now having many steps) by a smooth curve. The local gradient of this curve at any particular time will show the rate at which the population is increasing at that time. This rate, and therefore the gradient, may change over long periods of time. If the rate does not change, this smooth population curve will actually be a sloping straight line, whose gradient is the rate of population growth during that time. We will use such a sloping straight line below as such an approximation.

The school which I attended from 1943 to 1953 was The King's School in Grantham. One of the previous pupils (see below) was Sir Isaac Newton, whose influence there is always felt. He may be watching, in spirit, to see if we get the following calculation right.

The normal age range at The King's School is from 11 (entry) to 18 (leaving), divided into 7 successive age groups (I leave aside exceptional arrangements, such as third-year sixth-formers, for the purpose of this discussion). The School had 318 boys in 1951–2 when I was in the sixth form, but in 2011–12 this number had increased to 1020. So the School expanded by a factor of $1020/318 = 3.2$ within that period of 60 years.

We will assume that the same proportion (one seventh) of boys leaves each year, which means in particular that 45 (= 318/7) boys left in 1951–2, and 146 (= 1020/7) boys left in 2011–12. We will simplify our arithmetic by adopting the slight approximation that 50 boys left in 1952, and 150 boys left in 2012.

Then we can say that the mechanism in the intervening years which relates the number of intending leavers *n* (say) after the year *y* (say) is described by the equation

$$\frac{n - 50}{y - 1952} = \frac{150 - 50}{2012 - 1952} = \frac{100}{60} = \frac{5}{3} = 1.67.$$

Figure 17: Old School interior, King's School Grantham

Figure 18: Newton's name carved on an interior window-sill

In other words, the number n of boys who left in the year y is

$$n = 50 + 1.67(y - 1952).$$

[As a particular check, this formula can be seen to be consistent with our assumptions that $y = 1952$ implies $n = 50$, and $y = 2012$ implies $n = 150$.]

Now we can use this formula to work out that the *total* number N of leavers in all those 61 years since the summer of 1952 is the sum of the leavers in each individual year since then, namely

$$N = 50 + 51\frac{2}{3} + 53\frac{1}{3} + 55 + \ldots + 145 + 146\frac{2}{3} + 148\frac{1}{3} + 150.$$

This is the sum of what is called an *arithmetical progression* in which, as explained in the previous section, the difference between every pair of successive terms is the same, namely $1\frac{2}{3}$ in this example.

An apparent difficulty has arisen because we have been obliged to speak of $\frac{1}{3}$ of a boy, and $\frac{2}{3}$ of a boy. The mathematics does not 'know' that such fractions of a boy are not recognisable objects in School. But we do not need to tell it so, because the mathematics is clever enough to avoid such uncomfortable objects, as follows.

We write the above formula for N in the reverse order, to give

$$N = 150 + 148\frac{1}{3} + 146\frac{2}{3} + 145 + \ldots 55 + 53\frac{1}{3} + 51\frac{2}{3} + 50.$$

Now we add the two formulae vertically, adding corresponding terms in vertical pairs, to give

$$2N = 200 + 200 + 200 + 200 + \ldots + 200 + 200 + 200 + 200.$$

There are 61 terms in this series, so we deduce that

$$N = \frac{61 \times 200}{2} = 6100$$

is the total number of leavers in the 61 years between 1951–2 and 2011–12.

If all those leavers had paid a single ('one-off') £5 Life Membership fee to their Old Boys Society, that fund would have accrued £30 500 from those sources by now, ignoring any interest earned or disbursements paid. Perhaps all such leavers did not pay that subscription.

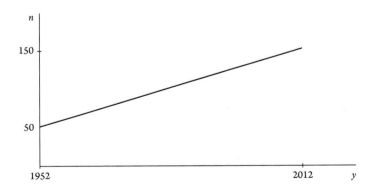

Figure 19: Increase of school leavers

Figure 20: Woolsthorpe Manor

The formula which we have deduced for n can be plotted as a straight line function of y. But this formula cannot realistically be extended very far *backwards* in time, because it predicts that the number of leavers was $n = 0$ when $0 = 50 + 1.67(y - 1952)$, i.e. *no* leavers when $y = 1922$. The School circumstances were obviously different before the 1939–45 War, and in particular when my father was a pupil there in 1920. The substantial Old School building in Grantham, which is still in regular current use, was built in 1528 by Bishop Fox of Winchester. The exterior and interior are shown in the three photographs kindly provided to me by the Head Master, Mr Frank Hedley. It was my job, as Head Boy in 1952–53, with the aid of other School Prefects, to get 300+ boys into that Old School building for the daily 'Assembly', and out again, in a disciplined way every weekday morning. But the School has 1300+ boys now, and so can no longer accommodate a whole-School assembly in the Old School building. The third photograph shows 'I. Newton' as one of many names carved (presumably by hammer and chisel) on interior window-sills in the Old School, probably two or three hundred years ago. Who carved that one? Of course the School also has other buildings now, on a large adjacent site. It is believed that there was a grammar school in Grantham in 1329. The School now groups boys in several so-called 'Houses', to one of which each boy is assigned for the extent of his School life, and Newton House is one of them.

The calculation in this section (without the photographs) was published in Mathematics Today **49**, 137, June 2013.

26 Isaac Newton

Sir Isaac Newton was one of the most famous and influential of English scientists. He was born in 1642 at Woolsthorpe-by-Colsterworth, near Grantham in Lincolnshire, and died in 1727. He is honoured, in particular, for his pioneering contributions to our understanding of the laws which govern the motion of bodies acted upon by gravitational forces, and for his study of optics; and for the development of the mathematics required to describe the effects observed in these topics. Did the apple fall to the ground, or did the Earth rise to meet it, or was it a bit of both? The pencil sketch (38 cm × 28 cm) of Newton's home at Woolsthorpe Manor was drawn by my sister Ann Garrard in 1956, and I am grateful to her for allowing me to reproduce it here.

   Newton's father died before he was born, but his mother was able to secure for him (as remarked above) an education at The King's School in Grantham, about eight miles to the north of his home. At first he travelled to School daily in a farm-cart. There have come down to us stories of him getting off on the way to School, and spending the day reading and writing mathematics by a roadside hedge, until the cart picked him up again on the way home. And later, when he lodged as a schoolboy in the town, he is said to have flown kites at night, from the upstairs window of his lodgings, and with lighted candles in them to surprise the populace.

   I have on my study wall the very old (black and white, apparently eighteenth century) engraved portrait of him (26 cm × 37 cm) reproduced here, which I bought in a Grantham antique shop for a few shillings in 1955. It shows him bewigged, and holding what may be the *Principia Mathematica*, his magnum opus. The engraving is evidently a copy of a (colour) painting of Newton done by John Vanderbank (1694–1739) in 1726, the year before Newton died at the age of 85. That painting is in the possession of The Royal Society (of which Newton was President from 1703 to 1727), to whom it was apparently presented by Martin Folkes, who was President of The Royal Society from 1741 to 1752. [J. Faber is named at the bottom of the engraving. Was he the engraver? Or the publisher, or the printer?]

27 Geometric Progression

A geometric progression is constructed when we begin with any number a (say), multiply it repeatedly (say n times) by another number r (the 'common ratio') and add the $n + 1$ numbers so formed. This process will construct a number which we will denote by

$$G = a + ar + ar^2 + ar^3 + \ldots + ar^n.$$

This is called a 'geometric series'. It is possible to construct a simpler version of this formula. To do this we multiply the series by r, giving

$$rG = ar + ar^2 + ar^3 + \ldots + ar^{n+1},$$

and then subtract G from rG. This gives

Figure 21: Sir Isaac Newton

$$(r-1)G = ar^{n+1} - a \text{ so that } G = \frac{a(r^{n+1} - 1)}{r - 1}$$

is the required compact (i.e. not a 'series') formula for G.

In the particular case when $r < 1$ we also have $r^{n+1} < 1$. In some cases, for small r and large n, r^{n+1} may be so small that it is negligible compared with 1, i.e. $r^{n+1} \ll 1$. Then, a reasonable approximation of the formula is just

$$G = \frac{a}{1 - r}.$$

This approximation for large n becomes exact as n tends to infinity whenever $r < 1$.

An application of this result in mechanics arises when we drop a ball onto a surface from which it bounces. It may bounce several times (e.g. a table tennis ball on a table), or only two or three times (e.g. a tennis ball on concrete), losing energy at each bounce. This poses the problem of finding the time taken for each ball to stop bouncing.

To solve this problem, suppose the ball has mass m and therefore weight $w = mg$. Here g is the acceleration which gravity imposes on any freely falling body when air resistance is neglected. Near the Earth's surface the value of g is about 32 feet per second every second, and it is the same for every body whatever its mass may be. If the ball begins to fall with zero initial speed, and it arrives at a lower level surface with speed v, the initial potential energy wh due to height h is converted into kinetic energy $\frac{1}{2}mv^2$ due to motion, so that $wh = \frac{1}{2}mv^2$. This gives a speed $v = \sqrt{2gh}$ of arrival at the surface. The time t taken for this first fall to the surface can be calculated, from the definition of acceleration g, to be $t = \frac{v}{h} = \sqrt{\frac{2g}{h}}$.

There is a loss of kinetic energy at the first, and each subsequent, bounce. This is described by supposing that the ratio of leaving speed to arriving speed is a constant e, say, called the 'coefficient of restitution', and such that $e < 1$. The value of this e will depend on the materials of which the ball and the surface are made.

The ball will therefore leave the surface after the first bounce with speed ev, and so the time which will elapse before the second bounce will be $\frac{2ev}{g}$. This fact comes from the formula 'speed difference = acceleration (g) multiplied by time', remembering that the ball goes up, and then comes down again (hence the factor 2).

Repeating this process for succeeding bounces, and adding the results, gives the total time of bouncing to be

$$T = \sqrt{\frac{2h}{g}}\left(1 + 2e + 2e^2 + 2e^3 + \dots\right) = \sqrt{\frac{2h}{g}}\left[1 + 2e\left(1 + e + e^2 + \dots\right)\right]$$

so that by summing this geometric series

$$T = \left(1 + \frac{2e}{1-e}\right)\sqrt{\frac{2h}{g}} = \left(\frac{1+e}{1-e}\right)\sqrt{\frac{2h}{g}}.$$

Here we have used the fact that each fall is preceded by a rise (hence the factor 2), except for the first fall, and we have used a formula for the infinite series which arises from the (in principle) infinite number of bounces (even though most of the later ones are too small to see).

For example, if the speed is halved by each bounce, so that $e = \frac{1}{2}$, the total time of bouncing is $3\sqrt{\frac{2h}{g}}$.

28 Zeno's Paradox

Zeno was living in the Greek town of Eleas in 450 B.C. He is credited with being the author of what is now called 'Zeno's Paradox'. It can be summarised by the assertion 'You cannot get Home'. What does this mean?

Suppose the distance from Here to Home is 1 unit. Then to get Home you must first go halfway. Then you must go half of the remaining half. That leaves the last quarter. Then you must go half of that. And so on.

Arithmetically you are accumulating increasingly small bits of the remainder of the journey, according to the sum

$$\frac{1}{2} + \frac{1}{4} + \frac{1}{8} + \frac{1}{16} + \frac{1}{32} + \frac{1}{64} + \dots \text{ and so on.}$$

It appears that we can keep repeating the procedure, but *never finish*. There is always half of a small remainder to go, so *we can never get there.*

What is the resolution of this difficulty? Introduce a label, say *D*, for the total distance to be travelled, so that we can *talk* about it efficiently. Then *D* is the *sum* of an *infinite* number of *decreasing* steps described by

$$D = \frac{1}{2} + \frac{1}{4} + \frac{1}{8} + \frac{1}{16} + \frac{1}{32} + \frac{1}{64} + \frac{1}{128} + \frac{1}{256} + \dots.$$

Can we actually determine *D*, now that we have described the problem more explicitly? Multiply both sides of this equation by 2, which gives

$$2D = 1 + \frac{1}{2} + \frac{1}{4} + \frac{1}{8} + \frac{1}{16} + \frac{1}{32} + \frac{1}{64} + \frac{1}{128} + \frac{1}{256} + \dots.$$

But the right-hand side can be seen from above to be $1 + D$, so that $2D = 1 + D$, and therefore $D = 1$.

This is exactly the distance from Here to Home, so you *do* get home, by taking an *infinite* number of *decreasing* steps. Thus we have added up the original infinite series for *D*.

We have summed a particular *convergent* geometric series or progression.

29 Big Birthday Problem

Two people of my acquaintance, and who met each other, are called Bunty, an old lady who was born on 10 November 1907, and Daniel, my grandson who was born on 7 January 2004. The Big Birthday Problem is to find out on which day their two ages add up to 100 years exactly. This Problem was posed on 22 April 2005 (called 'today' below).

First we work out how old they are today. We shall write *y* for years, *m* for months and *d* for days; and *B* for Bunty's age and *D* for Daniel's age.

$B = 2004y - 1907y = 97y$ approximately, and $B = 97y + 5m + 12d$ exactly.

$D = 2005y - 2004y = 1y$ approximately, and $D = 1y + 3m + 15d$ exactly.

So the sum of their ages today is $B + D = 98y + 8m + 27d$ exactly. Therefore, it will be approximately 7.5 months before their combined ages is 100 years (because the sum requires another 15 months approximately, to be equally divided between two people). But we want the exact solution.

As a preliminary calculation we work out their ages on Daniel's next birthday, 7 January 2006, on which $D = 2y$ exactly, and $B = 98y + 1m + 28d$. So $B + D = 100y + 1m + 28d$ on that day. This confirms that the Big Birthday must be before Daniel's next birthday.

Next we work out how many days there are between the two birthdays of 10 November and 7 January. It is

$$20 \text{ (in November)} + 31 (\text{December}) + 7 (\text{January}) = 58.$$

Now we introduce some 'algebra'. Suppose the Big Birthday is X days after Bunty's 98th birthday, and therefore $58 - X$ days before Daniel's 2nd birthday. The required fact can then be expressed as

$$(98y + Xd) + (2y - [58 - X]d) = 100y.$$

Can we solve this equation to find X? It simplifies to

$$100y + Xd - 58d + Xd = 100y, \text{ and then to } 2Xd = 58d, \text{ so that } X = 29.$$

Therefore, the Big Birthday is 29 days after Bunty's 98th birthday, and $58 - 29 = 29$ days before Daniel's 2nd birthday. So it will be on
9 December 2005.

On this day Bunty was 98 years + 29 days, and Daniel was 2 years − 29 days, so that the sum of their ages was 100 years exactly.

30 Sundays in February

Last Sunday (in 2004) was the fifth Sunday in February. How frequently does this happen? How can we think about this problem? The same question can be asked of any other particular day of the week. Only in a leap year can there be five February days with the same name. A leap year occurs in only one year in every four, provided the year is divisible by four, and provided it is not a century year.

So because there are seven days in a week, any February day with the same name will occur *five* times once in every $28 = 7 \times 4$ years, leaving aside the century year exception.

Put otherwise, $\frac{1}{4}$ of the years are leap years, and $\frac{1}{7}$ of the days are (for example) Sundays, so five Sundays in February occur in $\frac{1}{4} \times \frac{1}{7} = \frac{1}{28}$ of the years.

31 Think of a Number

Perform the following eight steps, with *any* starting number n, and the end result will always be 2.

Think of a number (*any* number, say 23), add 1 (to make 24), double it (48), take away 3 (45), add the number you first thought of (68), add 7 (75), divide by 3 (25), take away the number you first thought of (which finally delivers 2).

Why does this work? *Algebra* will explain it, as follows:

1. Think of a number: say n.
2. Add 1: $n + 1$.
3. Double it: $2(n + 1)$.
4. Take away 3: $2(n + 1) - 3 = 2n - 1$.
5. Add the number you first thought of: $2n - 1 + n = 3n - 1$.
6. Add 7: $3n - 1 + 7 = 3n + 6$.
7. Divide by 3: $(3n + 6)/3 = n + 2$.
8. Take away the number you first thought of: $n + 2 - n = 2$.

32 Handshaking

A group of p people meet, and each person shakes hands with every other person once. What is the number h of handshakes that will take place?

Conversely, if we know the number h of handshakes that take place, what is the number p of people participating?

We can start with the obvious observations that if $p = 2$ then $h = 1$, and if $h = 1$ then $p = 2$. Also, if $p = 3$ then $h = 3$; and if $h = 3$ then $p = 3$. But it would be false to jump to the conclusion that if $p = n$ then $h = n$, and vice versa, for $n > 3$.

A counter-example to that conclusion is provided by $p = 4$, which implies $h = 6$. The problem belongs to a class of problems called 'permutations and combinations'. There are 6 distinct ways of selecting pairs from 4 people. This can be illustrated as follows.

A stepped diagram can be drawn which illustrates how 4 people A, B, C, D can imply 6 handshakes AB, AC, AD, BC, BD, CD, as shown below.

AB

AC BC

AD BD CD

Then this diagram can be turned through half a full rotation and fitted next to itself to make a rectangle which repeats the information, and this tells us that for $p = 4$ people, $2h = p(p - 1)$ so that $h = 6$.

AB CD BD AD

AC BC BC AC

AD BD CD AB

Some further thinking tells us that this formula, for those particular values of p and h, generalises to give

$$h = \frac{1}{2}p(p-1)$$

handshakes between any number p people. This value of h is the sum of the first $p-1$ integers.

The converse question is: given the number h of handshakes, how many people p will participate? This requires the quadratic equation $p^2 - p - 2h = 0$ to have a positive integer number $p = \frac{1}{2}(1 \pm \sqrt{(1+8h)})$ of solutions. We can disregard the negative option, so we conclude that the number of people who participate in h handshakes is $\frac{1}{2}(1 + \sqrt{(1+8h)})$, provided $1 + 8h$ is a square number.

For $h = 0, 1, 3, 6, 10, 15, 21, \ldots$ the number of people required is $p = 1, 2, 3, 4, 5, 6, 7, \ldots$, respectively.

33 Losing Money to the Bank

On one occasion I discovered, by phoning the Bank, that the Bank would sell 1.3611 US dollars for 1 UK pound, and it would buy 1.5011 dollars per pound. Thus it would gain 1.5011 − 1.3611 = 0.14 dollars per pound.

In other words the Bank would make, over the counter from customers, a profit of 14 US cents for every UK pound it bought and sold. This is one of the ways by which the Bank acquires resources to pay their staff salaries and bonuses.

On a later day the exchange rates had changed. Now the bank would sell 1.3844 dollars for 1 pound, and it would buy 1.4520 dollars per pound.

Thus now the Bank gains 1.4520 − 1.3884 = 0.0736 dollars per pound. So within five weeks the Bank's profit on these transactions has been nearly halved, from 14 cents to 7.36 cents per pound.

The following is one associated problem which arises.

If I bought dollars five weeks ago, and came back from the United States having spent all but 100 dollars, how much do I lose by selling them back to the Bank now?

The mathematician in me finds it helpful to introduce some notation, of D for dollars and P for pounds. Then I can express the problem using equations, as follows.

I bought at the rate of $D = 1.3611P$, so that $D = 100$ cost $P = \frac{100}{1.3611} = 73.47$. Five weeks later I sell at the rate of $D = 1.4520P$. So for $D = 100$ I receive $P = \frac{100}{1.4572} = 68.87$. So my loss is 73.47 − 68.87 = 4.60 pounds when I return the 100 dollars.

34 Supermarket Offers: Deal or No Deal?

Having noticed that some of the price labelling in a local supermarket seemed to be ambiguous, I thought it would be instructive to ask myself questions about the actual meaning of the labelling, especially to the one-off shopper as distinct from the regular shopper.

I copied down the precise and full content of some of those labels in the supermarket, so that I could think about them later. ('Can I help you, Sir?' said the assistant. 'Not just now, thank you.')

In some cases genuine savings per item could clearly be made if more than one item were bought.

In other cases it could be concluded that no saving was actually being offered, although it seemed that the customer was being invited to conclude that a real saving *was* being offered. Some labels did no more than state a piece of arithmetic which the shopper already knew before entering the store. The pound symbol is customarily omitted in all the decimal labelling, and the customer is left to assume it.

The following five examples provide illustrations:

(a) Passion fruit was priced at 0.69 each, and the printed offer was '2 for 1.00'. Clearly this would deliver an actual saving of $2 \times 0.69 - 1.00 = 0.38$ per pair. *Deal.*

(b) Lychees were priced at 1.99 each, and the offer was '50 per cent extra' with 150 g deleted (a printed stroke through it) and 225 g printed next to it. This is *No Deal*: no actual saving was being offered, because the price is fixed (at '1.99 each'). The customer already knew, before entering the shop, that 225 is 150 per cent of 150. *No Deal.*

(c) Organic raspberries: 125 g was priced at 1.99, and the offer was 'Save 1.00' with, printed next to it, 2.99 and 1.99 both crossed out with a sloping line. This is *No Deal* because the price is fixed and the customer already knew that $2.99 - 1.99 = 1.00$ before entering the shop.

In this case there was extra printed information: 15.92 kg. This merely seems to mean that there is a notional background price of 1.592 per 100 g, which supposedly illuminates the already stated price of $1.99 = 1.25 \times 1.592$, but does not add anything new to it.

(d) Blackberries: 150 g were priced at 2.49, with an offer of '2 for 3.00'. This would deliver an actual saving of $2 \times 2.49 - 3.00 = 1.98$ per *pair* of boxes. *Deal.*

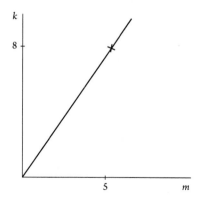

Figure 22: Kilometres and miles

In this case also there was extra printed information: 16.60 kg. This merely seemed to mean that there is a notional background price of 1.660 per 100 g, which supposedly illuminates the already stated price of $150 \times 1.660 = 2.49$, but does not add anything new to it. Instead it may merely confuse the shopper who would prefer to move on to the next item on his or her shopping list.

(e) Strasberries: 125 g were priced as 'Save 1.00' with a printed line through 3.99, and 2.99 printed next to it. This is *No Deal* because a price of 3.99 is not stated anywhere.

In this case too there was extra printed information: 23.92 kg. This merely seemed to mean that there is a notional background price of 2.392 per 100 g, which implies $1.25 \times 2.392 = 2.99$, but this is merely the already stated price for each packet of 125 g.

35 Imperial and Metric

The Imperial and metric scales of measurement provide an opportunity to discuss straight line graphs. For example, distance can be measured in miles or kilometres, and the relation between them is that 5 miles = 8 kilometres. Therefore 1 kilometre = $\frac{5}{8}$ of a mile. This relation can also be expressed as a straight line graph whose equation is

$$5k = 8m \text{ or } k = \frac{8}{5}m \text{ or } m = \frac{5}{8}k.$$

This $k(m)$ graph is only defined for positive k and m (unlike temperature graphs where negative Celsius temperatures mean freezing conditions). The $k(m)$ graph begins at $k = m = 0$ and passes through the point where $k = 160$ and $m = 100$.

Cricketers know that a good fast bowler can deliver the ball at a speed of 90 miles an hour (m.p.h.) or more, and the formula tells us that this is equivalent to $90 \times \frac{8}{5} = 144$ k.p.h. In some countries it has recently become customary to describe ball speeds in k.p.h. But anyone who has grown up using m.p.h. may have intrinsic difficulty in understanding that an orally quoted ball speed of, say, 128 k.p.h. means $\frac{5}{8} \times 128 = 5 \times 16 = 80$ m.p.h.

Imperial and metric measures of other quantities, such as weight (pounds and grams) or distance (yards and metres) offer many other opportunities for comparison. The origin of Imperial measures can often be found in practical usage (such as 22 yards = 1 chain used historically by surveyors; and was this the historical origin of the modern length of a cricket pitch, from over 200 years ago?), whereas metric measures are strongly based on powers of ten.

36 House Prices in Maidenhead

I noticed in the newspaper (this was in 2004) that Bill Gates, the Chairman of Microsoft, was said to be worth 27 billion US dollars. So I posed the problem: how many houses could Bill Gates buy in Maidenhead?

We first have to agree on the definition of 'billion'. I used to think it meant 'a million million', but more recently it seems to have become tacitly agreed, in the newspapers at least, that a billion is only 'a thousand million'. So we will use that definition.

Next we have to decide which currency to use, either US dollars ($) or UK pounds sterling (£). The current exchange rate was £1 = $1.8821, but to make the calculations easier I will use £1 = $2.

Therefore, if Bill Gates moved all his dollars into sterling at that rate, he would have

$$\$27\,000\,000\,000 = £13\,500\,000\,000.$$

So with this much sterling in his pocket, how many houses could Bill Gates buy in Maidenhead?

To decide this we must look at the price of houses, and to make the calculation easy I proposed that we work with an average house price of £250 000. This was immediately overruled by the children, now warming to the task, who asserted that to be realistic we would have to suppose an average price of at least £400 000 per house in Maidenhead.

However, to make a simple first calculation, the children agreed to go along with my first suggestion of £250 000 per house (albeit with the warning that these would be very small houses).

Thus we calculated that Bill Gates could buy

$$\frac{13500000000}{250000} = \frac{1350000}{25} = 5400$$

small houses in Maidenhead; but fewer if they were more expensive.

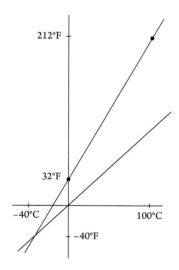

Figure 23: Fahrenheit and centigrade

37 Fahrenheit and Centigrade

The centigrade (C) and Fahrenheit (F) temperature scales are each based on two fixed points, namely the freezing point of water which is defined to be at 0°C or at 32°F (° means 'degrees'), and the boiling point of water which is defined to be at 100°C or 212°F (at a 'standard' atmospheric pressure).

Thus water becomes ice below 0°C and 32°F, and water becomes steam above 100°C and 212°F. Between these points, and on either side of them, both scales are divided up into *equally spaced* degrees, either 100°C or 180°F.

This 'equally spaced' property means that the relation between centigrade and Fahrenheit scales is a *linear* one, representable by the upper straight line graph as shown.

The gradient of this line is

$$\frac{212 - 32}{100 - 0} = \frac{180}{100} = 1.8.$$

This means that every 1 degree centigrade = 1.8 degrees Fahrenheit. 'Centigrade' means 100 spaces literally. In modern times it has become customary to replace the name centigrade by Celsius.

Anders Celsius (1701–1744) was a Swedish professor of astronomy at Uppsala University. He proposed the centigrade scale in 1742, and called it centigrade from the Latin for 'hundred steps'.

The straight line relation between the two scales can be described by the equation

$$\frac{F - 32}{C} = 1.8 \text{ or } F = 1.8C + 32 \text{ or } C = \frac{F - 32}{1.8}.$$

Normal human body temperature is 98.4°F and therefore $\frac{66.4}{1.8}°C = 36.9°C$, i.e. 98.4°F = 36.9°C.

The question can be asked: is there a temperature where both Fahrenheit and centigrade values are the same, i.e. where F = C? We can answer this in two ways, either graphically or analytically. Graphically , we can find out where the straight line F = C intersects the straight line F = 1.8C + 32.

This intersection is where F = C = –40 degrees.

This value is obtained by solving the equation C = 1.8C + 32, which gives

$$F = C = \frac{32}{1 - 1.8} = \frac{32}{-0.8} = -40. \text{ Thus } -40°C = -40°F.$$

38 Small Things

Names have evolved for decreasing units of measurement, in particular relating to time (epochs, millennia, centuries, years, days, hours, minutes, seconds), weight (kilograms, grams) and distance (kilometres, metres, centimetres, millimetres). I have chosen this spelling of metre (which not everyone would use) to distinguish it from that of meter for a measuring device.

It is worth mentioning the names of smaller units, and we use weight to illustrate these, referring only to metric names. Thus we have the following.

$$1 \text{ milligram} = \frac{1}{1000} \text{ gram, and } 1 \text{ microgram} = \frac{1}{1000} \text{ milligram} = \frac{1}{1000000} \text{ gram.}$$

As an example, in a packet of medicines, each tablet might have 125 micrograms of the active ingredient.

Smaller units are

$$1 \text{ nanogram} = \frac{1}{1000} \text{ microgram} = \frac{1}{10^9} \text{ gram} = 10^{-9} \text{ gram;}$$
$$1 \text{ picogram} = \frac{1}{1000} \text{ nanogram} = \frac{1}{10^{12}} \text{ gram} = 10^{-12} \text{ gram}$$

now using index notation for clarity, e.g. $10^{12} = 1000\,000\,000\,000$.
Even smaller units are

$$1 \text{ femtogram} = \frac{1}{1000} \text{ picogram} = \frac{1}{10^{15}} \text{ gram;}$$
$$1 \text{ altogram} = \frac{1}{1000} \text{ femtogram} = \frac{1}{10^{18}} \text{ gram;}$$
$$1 \text{ zeptogram} = \frac{1}{1000} \text{ altogram} = \frac{1}{10^{21}} \text{ gram;}$$
$$1 \text{ yoctogram} = \frac{1}{1000} \text{ zeptogram} = \frac{1}{10^{24}} \text{ gram.}$$

These are unfamiliar in common usage, of course; they do provide an illustration of the use of indices. In particular, we see that a negative index means a reciprocal, as in $10^{-1} = \frac{1}{10}$.

39 Leonardslee Cake

My wife and I visited Leonardslee Garden in Sussex, and at their teashop we bought a slice of circular sponge cake to share. I was given the task of cutting it into two equal halves. I chose to do this, not by the obvious method of cutting straight along the centre line of the slice towards the centre point of the cake, but instead by cutting perpendicularly across that centre line. This poses the problem of where to make that cut.

The shape of the original slice is called a segment of a circle (see diagram). It has a middle line CDE from the centre C of the circle. Perpendicular to this middle line is a chord FDG of the segment.

We begin by cutting along the chord FDG, to remove the circular edge of the cake, and that edge can be halved by a cut DE along the diameter of the cake.

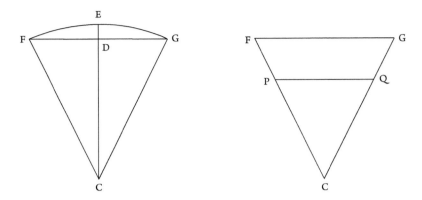

Figure 24: Original and triangular slices of mathematical cake

Figure 25: Original and triangular slices of the real cake

Denote the *lengths* of the base and height of the remaining triangle CFG by $B = FG$ and $H = DC$. Then the area of the triangle CFG is $\frac{1}{2}BH$, because two of them could be rearranged to make a rectangle.

Now suppose the one cut is PQ, parallel to FG, which leaves us with a smaller isosceles triangle PQC with base PQ = b (say) and height h. Its area will be $\frac{1}{2}bh$.

We require this area to be half the area of the larger triangle, so that $\frac{1}{2}hb = \frac{1}{2}(\frac{1}{2}HB)$. Therefore $HB = 2hb$ is the relation between heights and base lengths of the larger and smaller triangles. We also have the second fact that, because these two triangles are similar, we know that the ratios $\frac{b}{h} = \frac{B}{H}$ of base to height must be the same.

So we have the two equations

$$HB = 2hb \text{ (areas halved) and } Hb = hB \text{ (length ratios equal)}$$

which together imply $H^2 = 2h^2$. Therefore $H = \sqrt{2}h \approx 1.4h$ (and also $B = \sqrt{2}b$), and we make the cut to achieve $h \approx \frac{H}{1.4}$ as the height of the smaller triangle PQC.

'Did you work all that out in the cafe before cutting the cake?', asked one incredulous pupil with some emphasis.

40 Halving Areas

The Leonardslee cake calculation proved, by reasoning associated rather directly with an isosceles triangle, that areas of that shape can be halved if each linear dimension is reduced by the factor $\frac{1}{\sqrt{2}} \approx 0.7071067 \approx 0.71$.

Can it be proved that the same linear reduction factor will apply when an area of *any* shape is halved?

We can argue as follows. Any area measurement must be specified by a formula of the type $A = kBC$, where k is a constant and B and C are two linear dimensions.

Another area of the same shape but different values b and c for the corresponding linear dimensions will have the value $a = kbc$, with the same k as before.

Therefore the area reduction factor $\frac{a}{A} = \frac{b}{B}\frac{c}{C}$, i.e. it is the product of the linear reduction factors.

In particular, therefore, if the latter are to be the *same*, say $\frac{b}{B} = \frac{c}{C} = r$, then the area reduction factor will be r^2. Thus if, in particular, $r^2 = \frac{1}{2}$, then $r = \frac{1}{\sqrt{2}} = 0.71$.

41 Isosceles Tiling

An isosceles triangle has two internal angles the same (say B degrees for base, sometimes written as $B°$, but of course with a different meaning from temperature degrees) and one internal angle different (say $A°$ for apex). Therefore the lengths of the sides opposite each B are the same, but different from that opposite A when $A \neq B$. For example, we could have $B = 72°$ and $A = 36°$. This illustrates the property that the sum of the internal angles in any triangle is $180°$ ($= A + 2B$ in any isosceles triangle). The special isosceles triangle in which $A = B = 60°$ is called *equilateral*.

For any choice of B and $A = 180 - 2B$, we can join, on the opposite side of the base, another such triangle with the same dimensions; and then to each side of the second triangle, another pair of such triangles, as in Figure 26.

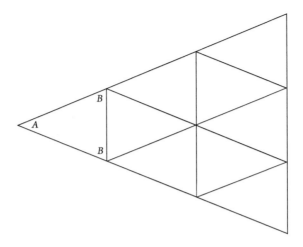

Figure 26: Isosceles stack of isosceles triangles

We then have four identical isosceles triangles, adjacent to each other in such a way that they make one larger isosceles triangle of the same shape, having the same internal angles, but being twice as high as the first one.

Now we can adjoin to the new base *two* triangles which are identical to the starting one, and *three* more in between those two. These five additional triangles complete a pattern of nine identical such triangles. Together, these nine comprise one larger isosceles triangle with the same angles *A*, *B*, *B* as the starting one. Such extensions can be continued indefinitely.

As well as the radial extension just described, we can add triangles cyclically by adjoining sides which are opposite the angle *B*. For example, if $A = 36°$ and we adjoin five such

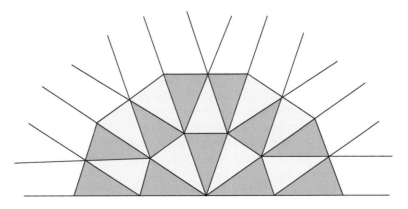

Figure 27: Isosceles tiling

triangles we shall obtain the fan-shaped pattern shown, with a straight-line base because of $5 \times 36 = 180$. Outward extension can be built along the five shorter sides and their corners. By so doing we shall achieve circuits consisting of 5, then 15, then 25, then 35 triangles, and so on.

Such patterns can be made out of slates and tiles. One such, with several circuits, could be seen on Open Days in Lord Carrington's garden at Bledlow in Buckinghamshire.

These tilings are examples of what may be called periodic tilings because they evidently contain a repeating (i.e. periodic) pattern. Non-periodic tilings have become of wide interest since their investigation by Sir Roger Penrose in the 1970s.

42 Roman Numerals

There are just seven Roman numerals, and their denary (base 10) equivalents are given in the following table.

Roman	I	V	X	L	C	D	M
Denary	1	5	10	50	100	500	1000

The method of constructing other numbers using Roman numerals is by *juxtaposition* (writing next to) as follows. A smaller number *before* a bigger one implies *subtraction*, such as IV = 5 – 1 = 4 and XL = 50 – 10 = 40. A smaller number *after* a bigger one implies *addition*, such as XV = 10 + 5 = 15 and MDL = 1000 + 500 + 50 = 1550. These two conventions can be combined or mixed, such as MCM = 1000 – 100 + 1000 = 1900.

Some more examples are XIX = 19, XVIII = 18, XXXXIX = 49 = IL, LXXI = 71, LXXIX = 79, LXXXXV = 95 = VC, CCCLXII = 362.

A notable awkward feature of the Roman system is that it has *no* symbol for zero. There is, however, one particular context in which this is *not* a handicap, and that is on clock-faces. These do not require a 'zero-hour', and that is one reason why it is so familiar to see Roman numerals on clock-faces for the numbers 1 = I, 2 = II, 3 = III, 4 = IIII = IV, 5 = V, 6 = VI, 7 = VII, 8 = VIII, 9 = IX, 10 = X, 11 = XI, 12 = XII.

43 New Money for Old

The British money system was changed overnight on 15 February 1971, to a so-called decimal coinage. The principal major unit of 'one pound' was retained, but this became subdivided in a different way from what it had been in the previous four hundred years. During that period the pound had been subdivided into 20 'shillings', and each shilling into 12 'pence' or 'pennies'. The symbols £, s and d were used for these pounds, shillings and pence, so that in particular £1 = 240d. This d may have been used to echo the old Roman *denarius*. The head of the current King or Queen was on one side (the 'obverse'), facing the opposite way to that of the previous monarch. The 'reverse' has displayed a variety of symbols, different for each denomination, and these symbols have changed with time.

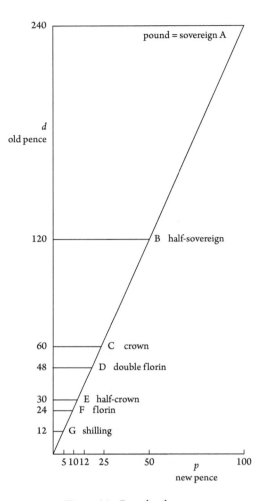

Figure 28: Pound and pence

The new subdivision created 'decimal currency', which means that 1 pound is now divided into 100 'new pence'. We write this fact as £1 = 100*p*. No other names for groups of new pence are used, in contrast to the pre-1971 period as described below. [One consequence, possibly unforeseen, of this innovation is that pricing of goods is now frequently expressed in a cumbersome fashion illustrated by the book which I bought for £16.99, instead of a straightforward £17 which I would have preferred.]

Market values before 1971, and especially before 1940, meant that £1 was a much larger amount of money then, in terms of buying power, than subsequently, so there was a practical need for many subdivisions of £1. For example, my father began his career as a school teacher in 1930, and he earned about £3 per week in those early years, in comparison with what would be in excess of £500 per week more recently.

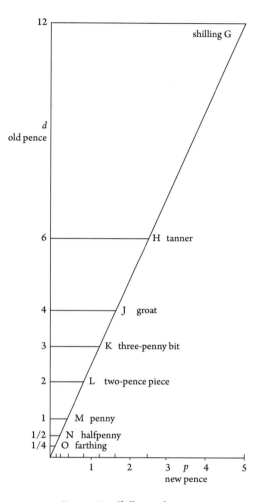

Figure 29: Shilling and pence

There was a large variety of designs on the reverse side of the pre-1971 coins. These included a wren on the farthing (1/4 *d*, denoted by O in Figure 29), which was withdrawn in 1960, a sailing ship on the halfpenny N (1/2 *d*), and Britannia seated with her shield on the penny M (1*d*), and a two-pence piece L (2*d*) in the 1800s. Those copper coins were accompanied at various times by silver coins such as the three-penny bit K (3*d*), the groat J (4*d*), the tanner H (6*d*), the shilling G (12*d*, also written 1/-), the florin F (2/-) which was a large and splendid silver coin in Queen Victoria's day, the half-crown E (2/6), the double florin D (4/-) and the silver crown C (5/-). All these were 'sterling silver' (92.5 per cent silver and 7.5 percent copper before 1920, after which the silver content was reduced to 50 per cent). This silver content of coins which have a silver colour has been zero since 1947.

Charles II 1679   George III 1770

George III 1797   George V 1911

Victoria 1887

Figure 30: Obverse sides of six old coins

Before the use of paper money there was also the gold half-sovereign B (10/-) and the gold sovereign A (20/- = £1). Some prices and prizes (for example in horse-racing) were (and still are) quoted in terms of the guinea (21/-).

There was also paper money as now, which in the 1940s (for example) included a particularly large and splendid five-pound note. Victorian coins, including copper pennies and halfpennies, were in common circulation until 1971.

All this variety, and with it the scope for a rich arithmetic of money, was swept away in 1971 and replaced by New Pence (p) and Pounds (still denoted by £) only, with 100p = £1.

This new decimal system makes calculation easier. A typical example is that £5.73 + £4.28 = £10.01. The old Imperial system had a mixture of 20 and 12 place values in columns.

Most of the Imperial coins were circular (except that silver and gold coins had milled edges, to prevent fraud by filing). Two of the most recent decimal coins (20p and 50p) are not circular, but instead their edges consist of seven circular segments whose centre of curvature is at the opposite circumference, so that they can still be accepted by slot machines designed for circular coins.

Charles II 1679   George III 1770

George III 1797   George V 1911

Victoria 1887

Figure 31: Reverse sides of six old coins

To understand readily the mathematics of the 1971 currency change, it is helpful to construct two straight-line graphs, each of which shows the relation between old pence and new pence, but for different ranges. One range shows the relation between (old) shillings and new pence, and thus for 5p up to 100p. The other range shows the relation between old pence and new pence, for 0d up to 12d, and thus for 0p up to 5p.

These graphs are similar to the Fahrenheit-centigrade graph shown previously, except that the latter does not pass through the origin.

It is still worth looking out for older coins. In March 2013, having just sat in my son's car on a parking patch of sun-dried flat mud by the road-side on Maidensgrove Common near Henley, I looked down as I was about to close the door, and I saw a circular shape. I thought to myself: 'That is a perfect circle'. I stretched down to put my finger-nail under it. It was a halfpenny of George III, dated 1780. I recognised it because I have two more in my collection. Presumably it had lain there unnoticed for perhaps 200 years. And in 2004, as I sat reading in my Reading garden, my wife came to me with something that she had found in our vegetable patch. 'What is this?', she said. 'Where did you find it?', I replied.

'I just dug it up' was her answer. It was a golden sovereign dated 1911, and unblemished as we saw when we brushed the mud off. That was the year the house was built. Perhaps the gardener had dropped his weekly wage? In 1950 I found an 1806 penny piece of George III in the garden of my home in Grantham. That garden may have been a field or pathway then; the house was built in 1930. My grandfather collected some old coins, and he gave a few to me, going back to the 1660s (Charles II). One of these is a splendid large Victorian silver double-florin (four shillings) dated 1887, commemorating Victoria's Golden Jubilee. As we said above, the advent of decimal coinage in 1971 (while I was abroad, researching mathematics for a year at The University of Wisconsin) removed, at a stroke, the historical presence of older coins from circulation.

The obverse and reverse photographs show a 1679 silver coin of Charles II, minted eight years before Sir Isaac Newton published his famous *Principia Mathematica*; and a tiny silver coin of that reign, perforated perhaps for a watch-chain; a 1770 George III halfpenny; a 1797 George III so-called 'cartwheel' twopenny piece, which has a diameter of 4 cm and is 5 mm thick, and is the largest coin ever used in England; the 1911 George V golden sovereign found in my garden; and the 1887 Victoria double-florin.

44 Distance Measures

An ancient (e.g. biblical) unit is

1 cubit = elbow to finger end.

Imperial units are also body-related, such as

12 inches = 1 foot, and 3 feet = 1 yard (stride).

A typical addition sum layout would be

yards	feet	inches
3	2	9
4	1	7
8	1	4

The metric units such as 100 centimetres = 1 metre (which is too long to stride) offer the decimal structure for addition.

45 Weight Measures

Imperial measures include

16 ounces = 1 pound, and 14 pounds = 1 stone.

These have a rich structure and a historical origin (e.g. related to a familiar object like a 'stone'). A typical addition sum may involve at least three columns thus:

stones	pounds	ounces
1	12	9
3	7	8
5	6	1

This can be contrasted with the decimal metric structure in which 100 grams = 1 kilogram, for which the typical addition sum is as simple as

$$1.94 + 0.27 = 2.21.$$

46 Perimeter-Diameter Ratios

For several closed geometrical figures it is informative to compare the measures of

perimeter p = distance all the way round the outside, and

diameter d = distance across from one side to an opposite one.

We measure d through the centre if there is a clearly definable one. Then it is of interest to compare the consequent ratios p/d. The following table is a sample of such conclusions.

polygon	p cm	d cm	p/d
thin rectangle	42	21	2
equilateral triangle	9	2.6	3.46
thicker rectangle	8.4	4	2.1
A5 rectangle	71.6	25.5	2.81
square	84	29.5	2.85
hexagon	15	5	3
dodecagon	15.6	5	3.12
circle	16.8	8.5	3.15

These are useful exercises in drawing and measuring. The fattest rectangle is the square. As the number of sides of a regular polygon increases, we see how the p/d ratio approaches that of the circle, whose exact value is π = 3.14159 approximately.

47 Fibonacci Numbers

There are various ways of introducing this topic. I did so by showing the pupils a 'Greek cauliflower', as it was called by the stallholder from whom I bought it for 80p in Woodley

Farmers' Market on 23 November 2008, two days before the lesson. I was in my 75th year, but I had never seen one before. The photograph shows it.

The Greek cauliflower exhibits a double spiral structure, with 8 spirals in one direction, and 13 spirals in the other direction. But when examined more closely, *each* of the florets which form the constituent parts of the two spirals *also* has the double spiral structure, again with 8 in one direction and 13 in the other. And then each of these smaller florets can themselves be seen to have the same (8,13) double spiral structure again. But the structure of the constituent parts at the next smaller scale turns out to be too small for that structure to be discerned.

This vegetable provides an opportunity to introduce, to 10-year-old pupils, a famous topic in mathematics. A sequence of integers can be constructed as follows. Start by writing down 1 and 1, add them together to get 2, construct a fourth number by $1 + 2 = 3$, a fifth by $2 + 3 = 5$, and continue to develop an infinite sequence by repeatedly adding the two previous members of the sequence. We get

$$1, 1, 2, 3, 5, 8, 13, 21, 34, 55, 89, \ldots$$

This sequence is called the Fibonacci sequence (or series). The historical origin of it is said to be its appearance in a book called *Liber Abaci*, published in Italy in the year 1202 A.D., and written by Fi Bonacci, the son of Bonacci.

A notation which is sometimes used for the sequence is

$$F_1 = 1, F_2 = 1, F_3 = 2, F_4 = 3, F_5 = 5, F_6 = 8, F_7 = 13, F_8 = 21, F_9 = 34 \text{ and so on.}$$

Figure 32: Greek cauliflower

This notation allows us to infer and display a simple formula

$$F_{n+2} = F_{n+1} + F_n$$

which describes how the sequence is constructed, for all positive integers $n \geq 1$ in the first place.

So the Greek cauliflower contains an example of an *adjacent* pair of Fibonacci numbers, namely $F_6 = 8$ and $F_7 = 13$. It is a remarkable fact that adjacent pairs of Fibonacci numbers occur in many situations in nature, as in this example and in others which I shall mention below.

Before that, however, we can notice firstly that the Fibonacci formula can be rearranged as

$$F_n = F_{n+2} - F_{n+1},$$

and secondly that it can be applied with non-positive integers $n \leq 0$. This allows us to define initially $F_0 = F_2 - F_1 = 1 - 1 = 0$, and then to extend the Fibonacci sequence by using $n = -1, -2, -3, \ldots$ to construct $F_{-1} = F_1 - F_0 = 1 - 0 = 1$, $F_{-2} = F_0 - F_{-1} = 0 - 1 = -1$, $F_{-3} = F_{-1} - F_{-2} = 1 - (-1) = 2$, $F_{-4} = F_{-2} - F_{-3} = -1 - 2 = -3$, $F_5 = 5$, $F_6 = -8$, $F_7 = 13$, $F_{-8} = -21$, $F_{-9} = 34$ and so on.

We notice that in this extension of the Fibonacci sequence the signs of the numbers alternate between + and –.

In summary, the Fibonacci sequence of numbers F_n when we include both positive and negative n is

$$\ldots 34, -21, 13, -8, 5, -3, 2, -1, 1, 0, 1, 1, 2, 3, 5, 8, 13, 21, 34, 55, 89, 144, 233, 377, 610, 987,$$
$$1597, 2584, 4181, \ldots$$

It does not seem to be customary to mention negative Fibonacci numbers in the literature.

A different way of expressing the construction of a Fibonacci sequence is as follows. Begin with any pair of numbers a and b, either of which may be positive or negative (or even equal, or zero). Construct, by adding (to the right), or subtracting (to the left), adjacent pairs. This procedure delivers the following sequence.

$$\ldots -8a + 5b, 5a - 3b, -3a + 2b, 2a - b, b - a, a, b, a + b, a + 2b, 2a + 3b, 3a + 5b, 5a + 8b, \ldots$$

In this construction a and/or b can also be allowed to contain, somewhere within it, the square root of -1, which is an 'imaginary' number usually denoted by i, or sometimes j. So this is a generalisation (apparently new) of the conventional statement of the Fibonacci sequence.

Another vegetable example of a Fibonacci pair is provided by sticks of Brussels sprouts. Of course these are also available in Farmers' Markets, and elsewhere. Examination of them reveals that the individual sprouts grow on the stalk in two different spiral arrangements.

Commonly there are 3 spirals anticlockwise moving upwards, and 5 spirals clockwise moving upwards. Thus they exhibit the Fibonacci pair $F_4 = 3$ and $F_5 = 5$, which is a different pair from that of the Greek cauliflower. But I have also seen the Brussels sprouts exhibiting the (2,3) pair.

The sunflower offers another well-known example of adjacent pairs of numbers in the Fibonacci sequence. After the flowers have fallen off we can see clearly that the arrangement of the seeds for the next year are in the form of spirals in opposite directions. Careful counting establishes that there are 34 spirals in one direction and 21 spirals in the other direction. This provides another example in nature of adjacent Fibonacci numbers. The photographs show a sunflower in full bloom, and then later when all the leaves and flowers have fallen off, and we can see the opposing Fibonacci spiral pattern. Not every flower is perfect, and some may contain isolated 'dislocations' in the pattern.

A pineapple also grows in a series of cells whose structure can be seen on the outside to be a pair of spirals in opposite directions. Examination of them shows there to be 8 spirals in one direction and 13 in the other direction. This is the same pair of Fibonacci numbers as we found for the Greek cauliflower.

Fir cones are another ubiquitous illustration of pairs of spiral growth in opposite direction. I have counted an (8,13) pair on a large fir cone found on Brownsea Island off Poole in Dorset. Smaller fir cones in Berkshire exhibit the Fibonacci pairs of (3,5) and (5,8) and (8,13).

We have described how the Fibonacci sequence can be extended algebraically to include negative numbers, and where they are present they alternate with positive numbers. This fact raises the question of how to interpret the negative Fibonacci numbers.

We can suggest here one possible viewpoint for interpreting the alternating sign in adjacent pairs of numbers in that part of the Fibonacci sequence which is 'to the left of zero'. The spirals always appear with alternating clockwise and anticlockwise sense. We can therefore (arbitrarily) assign the clockwise sense to be positive, and then the anticlockwise sense to be negative. If we do that we can offer the following summary of the observations which we have described.

Fir cones (5, –3), (–8, 5), (13, –8). The photograph shows 8 spirals going clockwise as seen from the bottom (with alternate spirals marked in white); and it shows 13 spirals going anticlockwise as seen from the bottom (with alternate spirals marked in green). In each direction, only alternate spirals are marked in order to see the distinction between the spirals more clearly. When there is an odd number of spirals thus marked, the first and last coloured ones are found to be adjacent, as shown in green.

Brussels sprouts (–3, 2), (5, –3). The photograph shows the latter type, having the steeper spirals repeated in 5s and the less steep spirals repeated in 3s.

Pineapple (13, –8). The photograph shows 13 spirals going clockwise as seen from the bottom (with yellow discs attached to alternate spirals); and it shows 8 spirals going anticlockwise as seen from the bottom (with red discs attached to alternate spirals). Again, in each direction only alternate spirals are marked in order to see the distinction between the spirals more clearly.

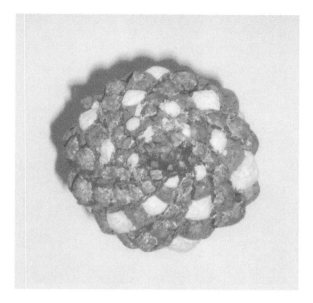

Figure 33: Fir cone

Figure 34: Brussels sprouts

Sunflower (34, –21).

Thus from 7 objects we have found 5 different Fibonacci pairs.

This topic offers a good exercise in learning to ask the ubiquitous scientific question: what is the pattern?

The *name* for patterning in plant development is *phyllotaxis*, and the sunflower (for example) displays what is called *double spiral phyllotaxis*.

Figure 35: Sunflower

48 Quadratic Equations and the Fibonacci Sequence

The equation $ax^2 + bx + c = 0$ in which a, b, c are given constants, either real or complex (i.e. possibly containing $\sqrt{-1} = i$), has two solutions $(-b + \sqrt{b^2 - 4ac})/2a = p$ (say) and $(-b - \sqrt{b^2 - 4ac})/2a = q$ (say) for x. This may be verified by substituting each, in turn, back into the left side of the equation. We can then use this pair of solutions to construct a sequence of numbers $(p^n - q^n)/(p - q)$ for $n = \ldots, -3, -2, -1, 0, 1, 2, 3, \ldots$. Substituting those values of n shows this sequence to be

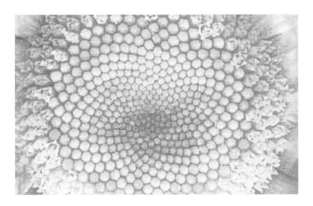

Figure 36: Sunflower spirals

Figure 37: Pineapple

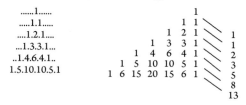

Figure 38: Pascal triangle, sheared, with Fibonacci numbers

$$\ldots - \frac{p^2 + pq + q^2}{p^3 q^3}, -\frac{p + q}{p^2 q^2}, -\frac{1}{pq}, 0, 1, p + q, p^2 + pq + q^2, \ldots .$$

In the case of the particular quadratic equation defined by the choices $a = 1, b = c = -1$, i.e. $x^2 - x - 1 = 0$, two solutions are

$$p = \frac{1 + \sqrt{5}}{2} \text{ and } q = \frac{1 - \sqrt{5}}{2}$$

and the central part of the sequence is $2, -1, 1, 0, 1, 1, 2$ which is the central part of the Fibonacci sequence.

49 Pascal's Triangle and the Fibonacci Sequence

Pascal's triangle is a pyramid of positive integers which is defined as follows. It has a sequence of 1's down the two sloping sides, with a single 1 at the apex, and a pair of 1's in the next (second) row on either side below the top 1. Each subsequent row is compiled by adding the two integers immediately above on either side of the number being calculated. The third, fourth, fifth, sixth,... rows are therefore 1 2 1, 1 3 3 1, 1 4 6 4 1, 1 5 10 10 5 1,... and so on. The integers in this pyramid are the same as the coefficients in the expansions $(a + b)^0 = 1$, $(a + b)^1 = a + b$, $(a + b)^2 = a^2 + 2ab + b^2$, $(a + b)^3 = a^3 + 3a^2b + 3ab^2 + b^3$, $(a + b)^4 = a^4 + 4a^3b + 6a^2b^2 + 4ab^3 + b^4$, ... and so on. These are called the binomial coefficients. The triangle is called Pascal's triangle (1653), after Blaise Pascal, 1623 – 1662, a French mathematician from Clermont and then Paris.

If we then shear the triangle horizontally (slide successive rows sideways) until one side (the right-hand side in the figure) is a vertical column of 1's, and then add the *diagonals* of this new triangle, we discover the remarkable fact that the results deliver the positive Fibonacci sequence $1, 1, 2, 3, 5, 8, 13,...$

These diagonal sums are $1 = 1, 1 = 1, 1 + 1 = 2, 2 + 1 = 3, 1 + 3 + 1 = 5, 1 + 4 + 3 = 8$, $1 + 5 + 6 + 1 = 13$ and so on.

The general binomial coefficients are those which appear in the expansion

$$(a + b)^n = a^n + na^{n-1}b + \frac{n(n - 1)}{1.2}a^{n-2}b^2 + \ldots + b^n.$$

50 Golden Ratio

This is a topic of wide interest, dating back to classical Greek times, and which is related to Fibonacci numbers. We listed these in the previous two sections, and we can now use them to define the ratios

$$R_n = \frac{F_n}{F_{(n+1)}} \text{ and } G_n = \frac{F_{(n+1)}}{F_n}$$

so that we obtain two more sequences of numbers, as follows.

n	-4	-3	-2	-1	0	1	2	3	4	5	6	7
R_n	-1.5	-2	-1	∞	0	1	0.5	0.666	0.6	0.625	0.615	0.619
G_n	-0.67	-0.5	-1	0	∞	1	2	1.5	1.67	1.6	1.625	1.615

It is evident from this table that both R_n and G_n oscillate as n increases through positive values, and also as n decreases through negative values.

As n increases through positive values, these oscillations diminish and tend to 'limiting' values as n tends to infinity which are

$$R_\infty = \frac{\sqrt{5} - 1}{2} \text{ and } G_\infty = \frac{\sqrt{5} + 1}{2}.$$

This value $\frac{\sqrt{5}+1}{2}$ is called the *Golden Ratio*. The Greeks knew it.

Because $\sqrt{5} = 2.23606679774\ldots$, the value of $G_\infty = 1.6180339887\ldots$ It appears in several contexts. Here we see that it is the limit of the ratio $\frac{F_{(n+1)}}{F_n}$ of successive Fibonacci numbers as n tends to ∞.

Five points A,B,C,D,E marked at every 72 degrees around a circle define the vertices of various possible figures having straight sides. Perhaps the most obvious would be the regular pentagon ABCDE obtainable by joining adjacent points. Another option would be the five-pointed star shown, which is called a pentagram, and is obtained by joining alternate points AC, BD, CE, DA and EB. There are also five isosceles triangles such as ACD which could be drawn.

Measurement shows that the ratio of the lengths of any long join (such as AC) to any short join (such as AB) is 1.62. This is very close to the golden ratio 1.6180339887.

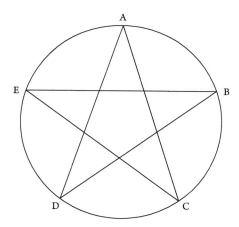

Figure 39: Pentagram

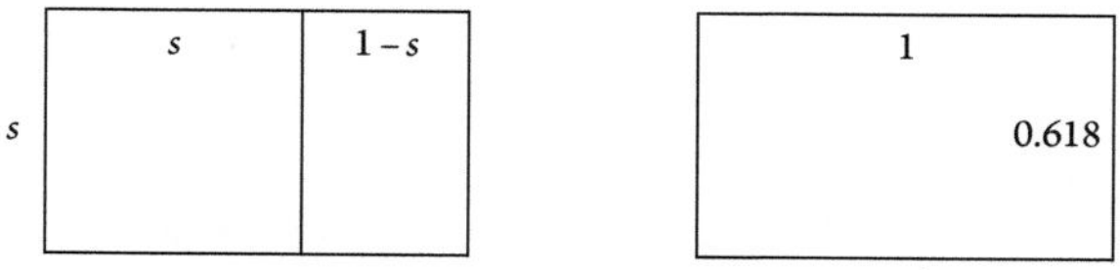

Figure 40: Golden ratio rectangles

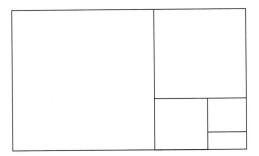

Figure 41: Nesting of golden sections

Golden ratios also occur in *rectangles*. Suppose we have a rectangle with long side of unit length, and with short side of length (say) $s < 1$. Divide the rectangle by a line parallel to the short side into a square with sides s and a rectangle with sides s and $1 - s$ as shown.

If we now require that the short/long side ratio be the same for the starting rectangle and the new smaller one, that means that s must satisfy the equation

$$\frac{s}{1} = \frac{1-s}{s} \text{ or } s^2 + s - 1 = 0.$$

This quadratic equation is the defining equation for the so-called 'Golden Section', and it can be rewritten as

$$\left(s + \frac{1}{2}\right)^2 = 1 + \frac{1}{4}.$$

This has two solutions

$$s = -\frac{1}{2} \pm \frac{\sqrt{5}}{2},$$

one positive and one negative. Only the positive solution has a meaning in defining the length of the shorter side as

$$s = \frac{\sqrt{5} - 1}{2} \approx 0.618.$$

A rectangle with these proportions is said to have the shape of the 'golden section', because it can be continuously subdivided into a square and a smaller rectangle, such that the sides of the latter have the 'golden ratio' short/long = 0.618 in every such subdivision. That is the 'golden' property. It reappears in other areas of mathematics. The Greeks are thought to have been pleased with the continuously repeating ratio property.

51 Pythagoras and the Fleet Air Arm

During my first visit to the Fleet Air Arm Museum at Yeovilton in Somerset in 2009, I found that in addition to the numerous historical aeroplanes, there was available an A5 leaflet, for visitors to take away, entitled 'How to calculate the distance to the horizon'. The recipe described there has the following three steps.

1. Estimate the height h of your eye above the sea surface.
2. Calculate $1.5h$ or $13h$, according to whether h is in feet or metres, respectively.
3. Take the square root $\sqrt{1.5h}$ or $\sqrt{13h}$, and these will be the distance to your horizon in miles or kilometres respectively.

Our first task is to justify these formulae.

What can be the basis of them? Something is being assumed about the shape of the Earth. Of course the starting point is to assume that the shape of the Earth is a sphere, at least approximately. The Earth is sometimes called a *geoid*. It is not an exact sphere, for two reasons. The first is a small-scale one, namely that there are valleys and mountains on land, and an undulating (waving) surface at sea which occupies the greater part of the Earth's surface.

The second reason is a large-scale one. The distance through the Earth from North Pole to South Pole is smaller than the diameter of the Equator, and this means that the shape of the Earth is what is called an *oblate spheroid*. Put otherwise, a cross-section through the Equator is a circle, but a cross-section through the Poles is a flattened oval. By contrast, the shape of a rugby ball is called a *prolate spheroid*, in which the diameter of the circular equator is less than the distance from pole to pole, as we mentioned in the section about 'spin-up'.

Having achieved this understanding of the basic real Earth geometry, we can then make an approximation that, for the particular purpose of calculating the distance to the horizon, we will assume that the Earth *is* a sphere, at least where the horizon is at sea. This could not be a reasonable approximation on land, because there might be a large hill directly in front of the observer.

So now we can draw Figure 42 to represent the assumed situation. This consists of (a) a circle of radius r to represent the surface of the Earth as a cross-section of a sphere through its centre C; (b) a radius from C to a point H representing the horizon on this surface as seen by the eye of an observer at O; (c) a tangent to the circle at that point H, to represent the line of sight OH to the horizon; and (d) another radius, extended by a distance h above the Earth's surface, so that h is the height of the observer above the surface. On this diagram S is the point at sea level which is directly below the eye of the observer.

Thus if the observer is standing on the sand with his eye at O, and with the tide lapping at his toes, then his feet will be at S. It would not be practicable for me to draw a scale drawing, because the Earth is so large compared with a man, and therefore r is really very much bigger compared with h than the diagram suggests (thus a mathematician would write $r \gg h$, in which the symbol $\gg$ stands for 'very much bigger than'). Having set the scene, we are now in a position to justify ('prove') the formulae quoted above for the distance to the horizon.

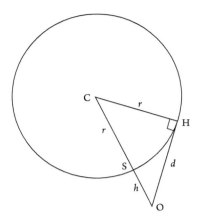

Figure 42: Circular cross-section of Earth, with observer

We need a version of Pythagoras' Theorem. What is that, and who was he? Pythagoras is believed to have been born on the Greek island of Samos in 569 B.C., and therefore 2585 years ago. Samos is in the Aegean Sea, part of the Mediterranean near to Turkey. Later he is thought to have moved to work in Sicily. Pythagoras is commonly thought to have been responsible for the proof that the sum of the squares of the two shorter sides of a right-angled triangle is equal to the square of the third side. This largest side is the one opposite the right angle, and it is called the hypotenuse.

This is one of the most famous theorems in mathematics. Laurence ('L.C.') Young expresses the view, in Section 8 of his learned book (1981), that 'it is fairly clear that in Babylon Pythagoras's theorem was familiar long before Pythagoras, but presumably without any proof'.

We shall prove it later, as we must, but first let us see how to use it in our situation. We have a triangle CHO with a right angle at the horizon point H. So Pythagoras' Theorem states that

$$CH^2 + HO^2 = CO^2$$

where $CH = r$, $HO = d$ and $CO = r + h$ stand for the lengths of the three sides. Thus we can write, as an alternative,

$$r^2 + d^2 = (r + h)^2.$$

The square of the hypotenuse on the right can be rewritten as $r^2 + 2rh + h^2$, so $r^2 + d^2 = r^2 + 2rh + h^2$ and therefore

$$d^2 = 2rh + h^2 = h(2r + h).$$

At this point our particular problem allows us to introduce an approximation. We know that the radius r of the earth is about 4000 miles, which is very much larger than the height of the observer (about 6 feet if he or she is on the beach, or perhaps 600 feet if he or she is on a hillside). So $h \ll 2r$ ($\ll$ means 'very much less than') and so we can *neglect* the height h of the observer compared with the diameter $2r$ of the Earth. This reduces the equation to

$$d^2 = 2rh.$$

This is an approximated version of Pythagoras' Theorem, with only a very small error. Taking square roots of both sides shows that the distance to the horizon is given by the simple formula

$$d = \sqrt{(2rh)}.$$

What does this predict when we put the actual numbers into the formula? Consulting an atlas tells us that the diameter of the Earth is $2r = 7926$ miles. For an observer on the beach whose eyes are 6 feet above the water, we must remember to use consistent units and therefore write $h = \frac{6}{5280}$ miles (because 1 mile = 5280 feet). Then the formula which we have proved above tells us that the distance to the horizon is

$$d = \sqrt{\frac{7926 \times 6}{5280}} = \sqrt{(1.5 \times 6)} = 3 \text{ miles}.$$

If an observer is on a hillside 600 feet above sea level, and can see the horizon at sea, then the distance of it will be $\sqrt{(1.5 \times 600)} = 30$ miles.

The *general* formula in terms of Imperial units is that when the eye of the observer is h feet above the local sea level, the distance to the observed horizon is

$$d = \sqrt{(1.5h)} \text{ miles}.$$

If we wish to use metric units, the fact that one kilometre is five-eighths of a mile gives the diameter of the Earth to be $2r = 7926 \times \frac{8}{5} = 12682$ kilometres. If we approximate this to be 13 000 kilometres, and if we use the notation that the eye of the observer is H metres and therefore $h = \frac{H}{1000}$ kilometres above sea level, the formula $d = \sqrt{(2rh)}$ gives the distance to the horizon to be

$$d = \sqrt{\left(13000 \times \frac{H}{1000}\right)} = \sqrt{(13H)} \text{ kilometres}.$$

For example, a tall man on the beach may have his eyes $H = 2$ metres above sea level, so that his horizon out to sea will be $\sqrt{26} \approx 5$ kilometres away.

We have used Pythagoras' Theorem, coupled with the approximation that the height of the eye is much less than the radius of the Earth to derive these approximate formulae

$$d = \sqrt{(1.5h)} \text{ miles} = \sqrt{(13H)} \text{ kilometres}.$$

Now we must prove Pythagoras' Theorem if we wish to be on firm ground.

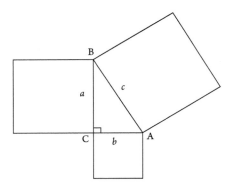

Figure 43: Squares on the sides of a right-angled triangle

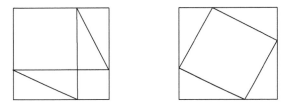

Figure 44: First proof of Pythagoras' Theorem

To do this we need to choose suitable notation. The choice of sensible notation is often an art in itself within mathematics, because a good notation can help to lead the discussion forward (and conversely, bad notation may hinder discussion). Suppose that A,B,C are the corners (vertices) of any right-angled triangle, with the right angle at C in the diagram. Denote the lengths of the sides by $a =$ BC, $b =$ CA and $c =$ AB. Then Pythagoras' Theorem says algebraically that

$$c^2 = a^2 + b^2,$$

or in geometrical language that the area of the square on the hypotenuse (shown in Figure 43) is equal to the sum of the areas of the squares on the other two sides.

To prove this construct the two larger squares of side length $a + b$ shown in the next diagram. Each of these squares has the same area $(a + b) \times (a + b)$. Each of them may be divided up, but in different ways, using four identical right-angled triangles in each, as shown. Each of these four triangles has a hypotenuse of length c, with the other two sides of lengths a (the longer one) and b (the shorter one).

The first subdivision contains two internal squares of total area $a^2 + b^2$, and four identical right-angled triangles each of area $\frac{1}{2}ab$. The second subdivision contains one internal square of area c^2, and four identical right-angled triangles each of area $\frac{1}{2}ab$. Because the original two large squares have the same area, we can conclude that the area of the single internal square on the right is the same as the sum of the pair of internal squares on the left, i.e.

$$c^2 = a^2 + b^2.$$

This completes the proof.

52 Fermat's Last Theorem

Pierre de Fermat, a French mathematician, claimed in 1636 to have proved that no generalisation of Pythagoras' Theorem is possible, in the sense that there are *no* trios a, b, c of positive integers for which

$$c^n = a^n + b^n$$

for any integer $n > 2$. In other words, $n = 1$ and $n = 2$ are the only cases for which such integer trios a, b, c can be found.

But Fermat gave no details of a proof to substantiate his assertion, and is quoted as saying tantalisingly, in the margin of his page, that 'this margin is too small to contain the proof'.

We had to wait until 1993 before Sir Andrew Wiles proved Fermat's claim, which is known as Fermat's Last Theorem, or *FLT*. Wiles' proof is neither brief nor elementary.

53 Another Proof of Pythagoras' Theorem

An alternative construction which proves Pythagoras' Theorem begins with two squares of sides a and $b < a$ placed next to each other, and with one side of each being collinear as shown. A point M on the common line PQ is selected so that it is at distances b and a from the corners P and Q of the larger and smaller squares respectively, in Figure 45.

Then lines MA and MB are drawn to the most distant corners A and B of the two squares. The triangles APM and MQB are congruent, and therefore the angle AMB must be a right angle. This is because it differs from the angle PMQ = 180° on a straight line by the sum 90° of the smaller pair of angles in either of those two congruent triangles.

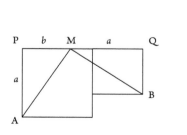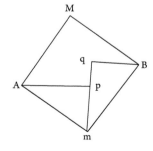

Figure 45: Second proof of Pythagoras' Theorem

This congruent property allows us to rearrange the diagram in the second version shown, in which the triangle APM goes to Apm, and the triangle BQM goes to Bqm.

The shapes and therefore their sizes have been preserved, and therefore their total area has not been changed. This area is $a^2 + b^2$ on the left, and c^2 on the right, where this length c = AM = MB = Am = mB is the length of the hypotenuse of the congruent right-angled triangles.

The idea of this proof is simple, but the description of it needs care. The idea of it has been attributed to Kelland in 1864 (by Bryant and Sangwin, 2008).

54 A Third Proof of Pythagoras' Theorem

We begin with any right-angled triangle ABC, having the right angle at the corner B (say), and side lengths BC = a, CA = b and AB = c (say). From B draw the perpendicular BH to H on the opposite side AC. This constructs two smaller right-angled triangles AHB and BHC, each similar to (having the same proportions as) ABC, as shown. Therefore the angle CAB (call it α) will be the same as the angle HBC. Because the proportions are the same, the areas of the three triangles CAB, BAH and CBH can be described by the three formulae $b^2 f(\alpha)$, $c^2 f(\alpha)$ and $a^2 f(\alpha)$, which have the dimensions of area, and in which we can use the same non-dimensional function $f(\alpha) \neq 0$ (even though it is unspecified explicitly). Therefore $b^2 f(\alpha) = c^2 f(\alpha) + a^2 f(\alpha)$, because the two interior triangles occupy the area of the large one. Hence $b^2 = c^2 + a^2$. I learned of this proof in a public lecture by J.R. Ockendon in 2011. I described it in the Autumn 2012 maths newsletter of The Royal Institution of Great Britain.

55 Another Application of Pythagoras' Theorem

The Ordnance Survey map of South Wales shows Ysgyryd Fawr, a mountain just north of Abergavenny, and called the Skirrid in English. I climbed it with friends, and I wanted to calculate the actual distance covered on the ground, using information about horizontal

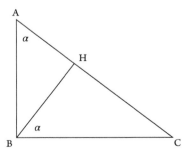

Figure 46: Third proof of Pythagoras' Theorem

distances and heights given on the map. Spot heights shown on the map at the beginning and end of the walk are 194 and 486 metres respectively, so the vertical part of the climb was 486 − 194 = 292 metres. The 1 kilometre grid lines on the map give the horizontal scale, from which I deduced, using a thread of cotton, that the horizontal component of the walk was almost exactly 2 kilometres = 2000 metres.

So the actual distance d metres travelled on the ground was certainly not less than the length of the hypotenuse of a right-angled triangle with horizontal side 2000 metres and vertical side 292 metres. Thus

$$d^2 \geq 2000^2 + 292^2 = 4\,085\,264$$

using Pythagoras' Theorem, so that $d \geq 2021$ metres approximately. This indicates that the actual distance travelled on the sloping ground was not less than 21 metres more than the horizontal separation of the top and bottom of the climb.

This is an underestimate of the actual distance trod because the path was uneven, so that we walked up and down in various places to make progress.

56 Pythagorean Triples

These are triples of whole numbers (also called integers) which have the property that the sum of the squares of the smaller two, which we can name x and y, is the square of the largest one z, so that $x^2 + y^2 = z^2$ is the defining property of the Pythagorean triple $x, y\, z$.

We can construct a table of the first few of these triples as follows:

x	3	5	8	7	9
y	4	12	15	24	40
z	5	13	17	25	41
$x^2 + y^2 = z^2$	25	169	289	625	1681

The table illustrates the arithmetic. Next we can illustrate the geometry by drawing one side of a triangle, measured with a ruler, and the other two sides to the intersection of arcs drawn by a pair of compasses set to the other two lengths; and then verifying that the angle opposite the longest side ('hypotenuse') is in fact a right angle (by measuring with a protractor). The reader may find it instructive to draw some of these diagrams, for example for the 3,4,5 triangle.

Such diagrams illustrate how a Pythagorean triple contains the lengths of the sides of a right-angled triangle.

There is scope with this topic for gaining experience in compass work. One pupil said: 'I think I need to go on a compass training course'.

The groundsman responsible for marking out the white lines on a football pitch can use a Pythagorean triple to help him get an accurate right angle at the corners. For example, he can take a rope 12 = 3 + 4 + 5 metres long, and tie a knot in it at distances of 3 metres and then 7 metres from the same end. If the rope is then laid on the ground with straight segments between the knots, and with corners at the two knots such that the two free ends meet, there has to be a right angle opposite the longest side. One of the corner flags can be located at that right angle. I have seen this procedure used.

Young (1981) remarks that it seems to have been known to Plato (427 B.C.–347 B.C.) that the pair of simultaneous equations

$$x^2 + y^2 = z^2, \quad x^3 + y^3 + z^3 = t^3$$

have one and only one set of positive integer solutions, namely

$$x = 3, \ y = 4, \ z = 5, \ t = 6,$$

for which the values of the two sides are 25 and 216 in the first and second members of the pair, respectively.

57 Nautical Notation

Sailors have good reasons for their own measurement systems. It is the case that the Earth's surface is not an exact sphere, although it is nearly so. It has a polar diameter of 7900 land miles (the length of the straight line through the Earth from North Pole to South Pole), but the diameter of the Equator is 7926 land miles. So, as we remarked previously, the Earth is an example of an *oblate spheroid*, being slightly flattened at the Poles.

If, however, we assume the Earth to be a perfect sphere, then a *great circle* is the cross-section on its surface by any plane through the centre. Then *one minute of arc* (equal to one sixtieth of a degree) on such a great circle defines one nautical mile, which is 6080 feet (and it is not the same as a *land mile* which is 5280 feet).

Sailors measure speed in *knots*, whose definition is that one knot is one nautical mile per hour, which is $\frac{608}{528} \approx 1.15$ land miles per hour.

The circumference of the Earth is $60 \times 360 = 21600$ nautical miles, which is $24873 = 21600 \times \frac{608}{528}$ land miles approximately.

58 Paper Sizes

The long and short side lengths L and S of a British standard (say A3) sheet of paper can be measured, and the ratio $\frac{L}{S}$ calculated. The paper can then be folded in half parallel to the short side, to make a smaller rectangle which is the size of the next smallest (then A4) sheet, whose long and short sides can again be measured. This procedure can then be repeated to get the next smallest rectangle which is called A5 size, then again to provide an A6 rectangle, and so on. The table shows the dimensions in millimetres, and the ratio $\frac{L}{S}$ is found to be $\sqrt{2} \approx 1.41$ at every stage. This last fact is the defining requirement for British paper sizes. Thus we obtain a specific illustration of an application of the equation

$$\frac{L}{S} = \frac{S}{\frac{1}{2}L} \text{ or } \frac{L}{S} = 2\frac{S}{L} \text{ or } L^2 = 2S^2.$$

L	S	$\frac{L}{S}$	Size name
419	297	1.41	A3
297	210	1.41	A4
210	149	1.41	A5
149	105	1.41	A6
105	74	1.41	A7

There is some interesting history behind the adoption of the $\sqrt{2}$ ratio for paper size. It was proposed by Professor Georg Lichtenberg in Germany in 1786, but it was not widely accepted. It was later proposed again, in 1922, by Dr Walter Porstmann, again in Germany, and then adopted as standard in that country. In 1975 it was adopted as an international standard by almost all countries, including the U.K., except in North America where a different ratio is still used (a standard page there measures 274 by 210 millimetres which is a ratio of 1.30, in contrast to the 1.41 which we mentioned above, for example with 297 by 210 millimetres for an A4 sheet).

59 Paper Sizes and an Infinite Sequence of Triangles

We begin with an A4 page, and halve it by drawing a horizontal line across the middle, parallel to the shorter side. Each half is A5. Halve the bottom half by drawing a vertical line down the middle. This creates two A6 shapes (as shown). Now halve the right-hand one by drawing a horizontal line across the middle, which creates two A7 shapes. Halve the bottom one by drawing a vertical line down its middle to create two A8 shapes. Halve the right-hand one by drawing a horizontal line across its middle, to create two A9 shapes. Continue this halving process until the rectangles are too small to work with.

Now draw the diagonal of the A4 page from the top left corner to the bottom right corner. This creates an infinite sequence of triangles, each occupying one quarter of the A5, A7, A9, A11,... rectangles.

We can prove, as follows, that the sum of the areas of this infinite number of triangles is $\frac{1}{6}$ of the area of the starting A4 page. This sum of the areas, in terms of fractions of the whole page, is

$$\frac{1}{4} \times \frac{1}{2} + \frac{1}{4} \times \frac{1}{8} + \frac{1}{4} \times \frac{1}{32} + \frac{1}{4} \times \frac{1}{128} + \dots$$
$$= \frac{1}{8} \times \left(1 + \frac{1}{4} + \frac{1}{16} + \frac{1}{64} + \dots\right).$$

If we use T to denote the sum

$$1 + \frac{1}{4} + \frac{1}{16} + \frac{1}{64} + \dots = T$$

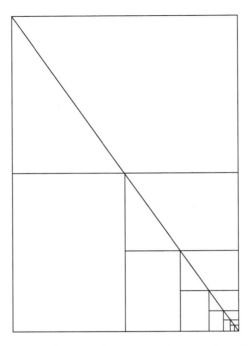

Figure 47: Sequence of paper sizes, and associated triangles

we see that

$$\frac{T}{4} = \frac{1}{4} + \frac{1}{16} + \frac{1}{64} + \ldots = T - 1$$

and therefore $1 = \frac{3T}{4}$ so that $T = \frac{4}{3}$.

Therefore the sum of the areas written above is

$$\frac{1}{8} \times T = \frac{1}{8} \times \frac{4}{3} = \frac{1}{6} \text{ of the area of the whole page.}$$

This is the same type of calculation as that which resolved Zeno's Paradox, in the sense that we are adding an infinite number of decreasing terms, but the terms are decreasing fast enough to make the answer finite.

60 Magic Squares

This is a brief introduction to a large topic.

The definition of a magic square is a square grid (which can be of any size) of numbers which has the following three properties. The sum of the numbers in every row is the same; and this is the same as the sum of the numbers in every column; and this is also the same as the sum of the numbers in the two main diagonals.

There is one case in which it is possible to *prove* the construction of a magic square in an elementary way. This is the case of a 2 × 2 square, which we can display as a *matrix*

a	b
c	d

consisting of four numbers *a, b, c, d* which are required to have the properties

$$a + b = c + d = a + c = b + d = a + d = b + c = m,$$

so that *m* is the common value of the six additions.

By subtracting pairs of these equations in three ways (first and third, or second and fourth; then first and fourth, or second and third; then first and sixth, or second and fifth) we reach the implication that

$$b - c = a - d = a - c = 0.$$

This *proves* that the stated six summation properties imply that all four numbers in the starting square *have to be the same*, namely

$$a = b = c = d.$$

That fact might have been guessed, but we have shown how to *prove* the result.

An example is $a = b = c = d = 7$, giving $m = 14$. There is a solution of this type for 2 × 2 squares with *any* even number $2n$ as the sum, so that $a = b = c = d = n$ is every entry in the 2 × 2 grid, which therefore becomes

n	n
n	n

.

An example is

7	7
7	7

.

As it turns out, the same problem becomes much more difficult when we increase the size of the square. The next one would be a 3 × 3 square with nine numbers, required to have the same sum in the three horizontal, the three vertical and the two diagonal directions. The individual entries in the 3 × 3 square no longer have to be all the same, and an example of such a solution is

2	7	6
9	5	1
4	3	8

.

The numbers in this 3 × 3 magic square are all different, instead of being necessarily all the same. Their sum in each of the horizontal, vertical and diagonal directions is 15 in this example.

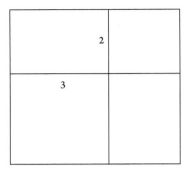

Figure 48: Binomial square

61 Binomial Squares

This name is sometimes given to a situation where a square of side lengths $a + b$ is subdivided into two smaller squares with sides a and b, and one common corner, and two rectangles with sides a and b (as shown with $a = 3$ and $b = 2$). Then the total area of the whole is $(a + b)^2 = a^2 + b^2 + 2ab = $ two different squares + two equal rectangles.

62 Some Special Squares

A square of side 2 units, when drawn entirely within a square of side 3 units, but not necessarily with sides parallel to it, will leave an area between them of $3^2 - 2^2 = 5$ square units, and we notice that $3 + 2 = 5$.

Similar examples are that

$$4^2 - 3^2 = 7 \text{ has the property } 4 + 3 = 7, \text{ and}$$
$$5^2 - 4^2 = 9 \text{ for which } 5 + 4 = 9.$$

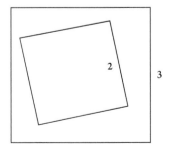

Figure 49: Example of some special squares

Are these three facts just isolated coincidences, or is there some underlying general property that we should notice?

We explore this with *any* two starting numbers x and y, for which

$$x^2 - y^2 = (x + y)(x - y).$$

From this *general* formula we can see at once that

$$x^2 - y^2 = x + y \text{ if and only if } x - y = 1.$$

We *do* have this property in the three examples with which we started, namely $3 - 2 = 4 - 3 = 5 - 4 = 1$, but it does *not* apply in other cases where $x - y \neq 1$.

63 The Nine-Point Circle

This result is a property of *any* triangle, so it is best to begin the discussion by excluding the following special triangles. Thus we avoid an equilateral triangle (all sides the same length), an isosceles triangle (just two sides the same length) and a right-angled triangle (in which one interior angle is 90°). We also avoid triangles which contain an obtuse internal angle (more than 90°), to make the drawing easier.

The Nine-Point Circle Theorem states that in any triangle with vertices A, B, C (see diagram)

(a) the three midpoints E, F, G of the sides BC, CA, AB, respectively;

(b) the three feet H, J, K of the altitudes AH, BJ, CK (so that AH is perpendicular to BC at H, i.e. AH is the height of A above BC, and so on); and

(c) the midpoints R, S, T of the lines joining the vertices A, B, C to the orthocentre O (where the altitudes intersect)

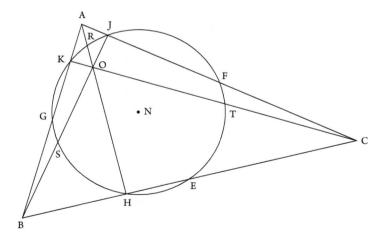

Figure 50: Nine-point circle

all lie on a single circle. This result dates from about 1822. The diagram shows the centre N of this circle, which lies at the intersection of the perpendicular bisectors of any pair of chords, such as SE and GK for example.

Proof:
The nine-point circle property can be proved as follows. Since AR = RO and AG = GB, GR is parallel to BO. Also, since BG = GA and BE = EC, EG is parallel to AC. But BO is perpendicular to AC, so that GR is perpendicular to EG. Hence the angle EGR is a right angle.

Similarly EFR is a right angle. But EHR is a right angle, and therefore the circle on ER as diameter passes through FGH. Therefore H and R lie on the circle EGF.

Similarly J,S and K,T lie on the circle EGF. Therefore the nine points EGF, HGK, RST all lie on the same circle. Q.E.D.

Moreover each side of the triangle EFG is half of the corresponding side of the triangle ABC.

Proofs are found in many books. One is Theorem 17 in the 1952 edition of *Modern Geometry* by C.V. Durell, first published by Macmillan (London) in 1920.

64 The Thirteen-Point Circle

It can be proved that the nine-point circle also passes through four more different points described below, so that it is really a thirteen-point circle. David Wells says, on page 76 of his 1991 book called *The Penguin Dictionary of Curious and Interesting Geometry*, that Karl Feurbach proved in 1822 (in Germany) that the nine-point circle also touches the inscribed circle of the triangle and each of the three exscribed circles of the triangle. Thus in that sense the nine-point circle is also a thirteen-point circle. An exscribed circle touches one side of the triangle between two vertices, and the other two sides when these are extended ('produced') beyond the vertices which they join.

65 Cardioid

This is a famous topic, which is a good illustration of a curve drawn as an *envelope* of straight lines.

Draw a circle, with the compasses, and not too small. Mark 36 points on it, at every 10 degrees, using the protractor. Label these points 1, 2, ..., 36 on a first cycle. Now use straight lines to join each point on the *right*-hand semicircle (in the diagram) to its 'double', so that for example $1 \rightarrow 2, 2 \rightarrow 4$ and so on, including for example $5 \rightarrow 10, 14 \rightarrow 28$ and $18 \rightarrow 36$.

Next, introduce a second cycle of labelling 37 to 72 and continue the joining process with straight lines from each point to its double. This time the outcome will be a joining of the *left*-hand semi-circle to its double, so that for example $19 \rightarrow 38 = 36 + 2, 25 \rightarrow 50 = 36 + 14, 32 \rightarrow 64 = 36 + 28$ and so on.

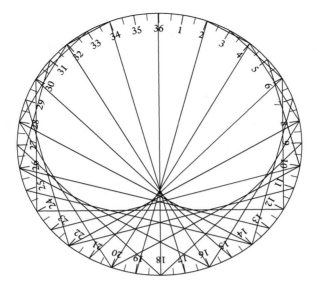

Figure 51: Cardioid

These two constructions, of a set of 18 + 18 straight lines, will combine to *envelop* a curve as shown in the diagram, and this curve is called a *cardioid* ('heart-shaped'). Unsurprisingly I have also heard it called (*sotto voce* in the classroom) 'bum-shaped', but at least this was evidence that the shape had been noticed.

Another convincing demonstration of the cardioid curve can be achieved on the surface of a quarter-full cup of coffee, with light provided by a beam of sunlight shining obliquely onto the inside of the cup. This parallel light beam is reflected off the interior side-wall of the cup onto the coffee surface, where the cardioid can be seen. Other light sources which demonstrate the same effect if the sun is not shining include a classroom light-bulb or a hand-torch.

66 Irregular Hexagons and Pappus' Theorem

A *hexagon* is a polygon in the plane obtained by joining six given points, called vertices, by straight lines. In the most familiar case none of these straight lines ('sides') cross each other.

However, we can allow the lines to cross each other, at points which we will then call self-intersections of the hexagon. These self-intersections are *not* the vertices.

This possibility invites an instructive exercise in classifying such types of hexagon, as in the diagram. None of the hexagons need be regular in the sense of having sides which are all the same length, although they could be so.

The diagrams show how we could have (a) no self-intersections (the 'open' hexagon), (b) one self-intersection ('scissors'), (c) two self-intersections, (d) three self-intersections, (e) four self-intersections, (f) five self-intersections. Obviously there could be more self-intersections.

When we come to seven self-intersections, there is a famous result called Pappus' Theorem (300 A.D.) associated with a particular such hexagon, as follows. Start by drawing two non-parallel straight lines, and mark three points on each. Then construct a hexagon with sides joining pairs of those six points, where the two points in each such pair are (see diagram) not on the same starting line.

Pappus' Theorem states that the central trio of these seven self-intersections all lie on a straight line.

Much more about this can be read in the 'Geometry' book by Brannan, Esplen and Gray (1999).

67 Regular Hexagons

This is an elementary topic which provides a worthwhile exercise in the use of a pair of compasses. The pencil needs to be no longer than the length of the pointed arm of the compasses, so that the pencil point is set next to the steel point of the instrument when in

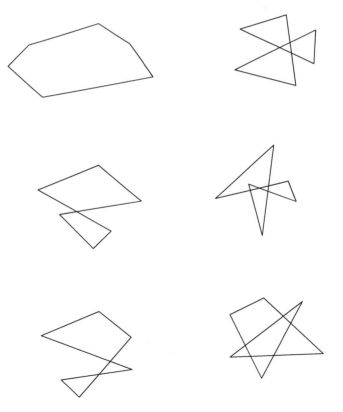

Figure 52: Hexagons with up to five intersections

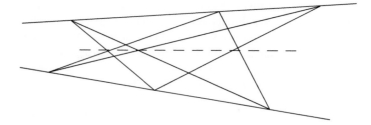

Figure 53: Example of Pappus' Theorem

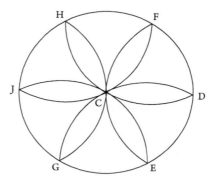

Figure 54: Flower pattern

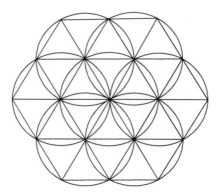

Figure 55: Honeycomb

the closed position. We open the compasses to any desired measurement (to be the radius) between the compass point and the pencil point, and we draw a circle with centre C.

Then, with the compass point anywhere on this circle (say D in the diagram), and with the same radius, we draw another circular arc which will pass through the centre of the circle and meet it at E and F. Next, with the compass point at E and F in turn, and the same radius, we draw two more circular arcs through the centre. They will both meet the circle at D, and each will meet the circle at other points G and H respectively. Next, with the compass point at G we draw another arc through E and C to meet the circle again at J; and with the compass point at H we draw another arc through F and C which will also arrive at J. Finally, with the compass point at J, we draw another arc GCH. The final result is a six-point 'flower pattern'.

A larger version of the same diagram can be drawn on an asphalt playground by a pair of pupils using a piece of string tied to a fixed point C at one end, and to a piece of chalk held by a pupil at the other end. If all the circular arcs like GCD are extended to full circles, we obtain an extended tessellation of the playground in the form of a hexagonally symmetric pattern of circles.

The six-point symmetry of DEGJHF with C means that when we join the points by straight lines we get a hexagonally symmetrical grid like a honeycomb, each containing six equilateral triangles, with all internal angles being 60°.

68 The Rugby Riddle

This is a problem which I was provoked to clarify by an ambiguous presentation on television of the dimensions of a rugby field. The following remarks are a simplification of some of the points contained in my article called 'A Rugby Riddle', published on pages 126–128 of *Mathematics Today*, Volume 41, in August 2005.

When I broached this topic with the question 'Who plays rugby?' to a class of 5 girls and 3 boys, all 10-year-olds, the immediate reply was 'We all do!'. Although one girl said there was no television in her house, another said that it was not possible to watch anything else when rugby was on, because that took precedence.

When a try is scored (5 points), at point T in the diagram, the scoring side has the right to attempt to 'convert' it into a 'goal' (2 more points) by kicking the ball over the bar between the posts A and B, from any point K it chooses which is on a line parallel to the touchline, and starting from the touchdown point T of the try, as shown in the diagram. Then the kicker has to kick between the lines KA and KB, and over the bar which joins the posts at A and B. This much geometry is not ambiguous.

However, when a try was scored, and before the kick was taken, numerical information appeared on the television screen, represented by the following example: 'distance' 41.9 metres, 'angle' 23 degrees and 'apparent goal width' 5.1 metres. In later matches, the last item was changed to 'visible goal width' expressed as a percentage (of what, was not stated).

This information is ambiguous, and requires clarification which a competent geometer would hope to be able to provide. Discussion of the ambiguities is instructive for ten-year-olds, even if the underlying trigonometry required for a fully explicit calculation is not yet within their knowledge.

For example, 'angle' must be AKB, i.e. that at K between the two lines KA and KB to the goalposts. That much is clear.

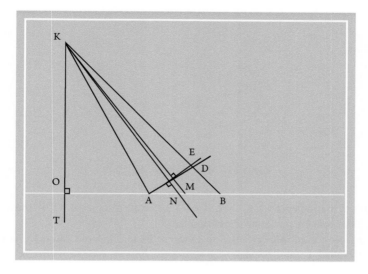

Figure 56: Rugby goal target

'Distance' must be that from K to some point on AB, but that point must be (and was not) specified. For example, this target point on AB might be either the midpoint (say M) of AB, or it might be where the bisector of the angle AKB meets AB (say at N, which is not the same as M); or indeed at any other point within AB which the kicker might choose to target.

'Apparent' or 'visible' goal width might mean, for example (although there is room for varied opinion), the length of the line from the nearest goalpost A, to the further direction KB, which is perpendicular to *one* of the 'distance' directions specified in the previous paragraph (e.g. AE perpendicular to KM, or AD perpendicular to KN). These 'goal widths' are different, in every case except when the touchdown point of the try is symmetrically behind the posts (in which case TK will be the perpendicular bisector of AB).

Trigonometry can be applied (as I did in my quoted 2005 article) to provide formulae for the apparent widths AD and AE observed by the kicker, although that calculation might be beyond ten-year-olds, as stated above.

A further question arises. Given the touchdown point T, how far out should the kicking point K be chosen to maximise the angle AKB subtended by the goalposts? Trigonometry can again be applied, to prove that this widest visible angle is obtained, to within a close approximation, if KM makes an angle of 45° with the goal-line AB.

It is worth quoting here, but without the proofs, the trigonometrical formulae for target widths quoted above. The apparent width between the posts when the kicker looks along the angle bisector AN is

$$AD = 2\sin\frac{1}{2}(\beta - \alpha)(k^2 + a^2)^{\frac{1}{2}}$$

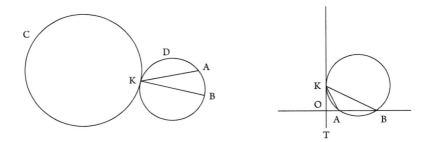

Figure 57: Maximisation of rugby target

in terms of the angles α = TKA and β = TKB and the distances k = OK and a = OA (O is where TK meets the touchline at right angles).

The apparent width between the posts when the kicker looks towards the midpoint between the posts, i.e. along KM, is

$$AE = \frac{(b-a)k(k^2+m^2)^{\frac{1}{2}}}{bm+k^2}$$

in terms of the known distances a = OA, b = OB, k = OK and m = OM = $\frac{1}{2}(a+b)$.

Publication of my article about The Rugby Riddle prompted Dr John Preater of Keele University to offer (*Mathematics Today* **41**, 163) the following observation. Suppose a given circle C has two fixed points A and B outside it. A point K can move on C, and we wish to maximise the angle AKB. The solution requires the construction of the smallest circle D that touches C (at K) and passes through A and B.

In the rugby problem, when the touchdown point T is outside the posts A and B, the angle AKB is maximised when K lies on a circle KAB which is tangential to the line TOK at K; so K is found by constructing a circle through the posts A and B and which touches TOK at K. More information is available in *Mathematics Today* **41**, 196.

69 Family Trees in People and Bees

A discussion of family trees and how to draw them is always a motive topic with a group of ten-year-olds. Immediately they foresee possible complications, for example what to do about people who are dead, what to do about step-children, or people who have been married twice.

If we concentrate on immediate blood relatives, that in itself is a worthwhile exercise in simplification. Ruthless simplification provides just four grandparents (GF$_1$, GM$_1$, GF$_2$, GM$_2$), two parents (F$_1$ and M$_2$) and one child (C), as shown in the family tree in people.

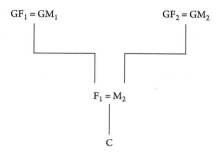

Figure 58: Family tree in people

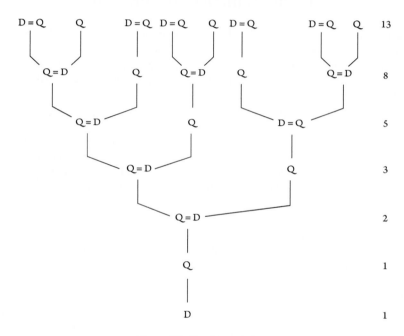

Figure 59: Family tree in bees

Thus one child will have had two parents, four grandparents, eight great-grandparents and therefore 2^n $[(n-2) \times$ great$]$-grandparents (so that, for example, there will have been 32 great-great-great-grandparents). As soon as we get away from the simplest situations, we can encounter convoluted trees, and it happens, for example because of siblings.

With bees, it is different from people. Just as there are two sorts of people, men and women, there are two sorts of bees, called queens and drones. Each person has two sorts of parent, one male and one female.

But not all bees have two parents. Queens have two, a queen and a drone, but drones have only one parent, a queen. So the family tree of a seventh-generation drone would be constructed as follows. The numbers on the right in the diagram are the count of the number of bees in that generation.

Very remarkably, we see from the numbers on the right of the family tree in bees that the numbers of bees in each previous generation are described by the Fibonacci sequence 1, 1, 2, 3, 5, 8, 13, ... So we can use that observation to predict the numbers of bees to be expected in earlier generations, which will be 21, 34, 55, 89,... and so on, without having to draw the diagram.

70 The Tethered Goat Problem

We have a goat in a square field, and we have a piece of rope whose length is half that of the side of the field. We also have a peg which can be fixed at any point within the field. The goat can be tethered by attaching the two ends of the rope to its collar and to the peg.

The problem is: how many different positions of the peg will be needed so that the goat can graze the whole field?

This can be treated as a practical exercise in the use of a straight edge and a pair of compasses.

So first we have to draw a square. One side can be easily drawn to any desired length, with the straight edge. (We choose to avoid using a ruler as a measuring instrument.) Next we use the compasses to erect perpendiculars at the two ends of the side just drawn, and with lengths equal to the first side. The other ends of these perpendiculars can then be joined to form a square.

The centre of the field can then be found as the intersection of the diagonals from the corners (or as the intersection of perpendicular bisectors of adjacent sides) of the square.

The goat is then introduced into the field, and tethered first with a peg at the centre, and then later with a peg at each one of the four corners in turn. This five-point symmetrical solution would certainly allow the goat to graze the whole field.

But it is also possible to find four-point solutions. For example, two are shown: one symmetrical and one unsymmetrical. In the unsymmetrical one, two peg positions lie on one diagonal, each being a quarter of the way in from opposite corners. Then the remaining two centres can be at the other two corners. The goat can cover the field from these four positions. In the symmetrical solution, all centres are on the diagonals. The diagrams illustrate these solutions.

71 Fencing the Bulls

This is the problem of finding the minimum number of straight fences that are needed to separate several bulls in a convex field, in the case when each fence goes to the boundary of the field, and the fences can intersect. 'Convex' means that the boundary is not indented.

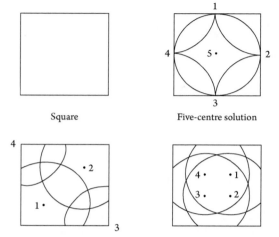

Figure 60: Tethered goat solutions

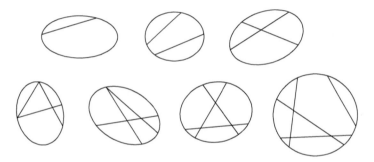

Figure 61: Fences for 2,3,4 and 5,6,7,8 bulls

It is helpful to begin with the simplest special cases, which are that 2 bulls need 1 fence, 3 bulls need 2 fences and 4 bulls need 2 intersecting fences, as shown. Continuing, we can make a table as follows.

bulls	2	3	4	5	6	7	8
fences	1	2	2	3	3	3	4

We have chosen to avoid constructing special situations such as three fences passing through a single point.

The problem could be turned on its head by asking a different problem, as follows.

For a given number of fences (lines), what is the maximum number of bulls (spaces) that they can enclose? The tabulated solution for this can be seen to be the following.

fences (lines)	1	2	3	4	5
maximum bulls (spaces)	2	4	7	11	16

Observation of how this table evolves shows that when the number of fences is increased from n to $n + 1$, this increases the maximum number of spaces (bulls) by $n + 1$. For example, when the maximum number of fences is increased from 3 to 4, this increases the maximum number of spaces by $4 = 11 - 7$.

Thus we can now extend the table without having to draw diagrams of the fences. The continuation is as follows.

lines (fences)	4	5	6	7	8
spaces (maximum bulls)	11	16	22	29	37

It can be inferred from the evolution implied in the discussion that the maximum number B_n of bulls (spaces) which can be enclosed by n fences (lines) is

$$B_n = \frac{1}{2}n(n + 1) + 1.$$

As a check we can see that this formula provides the table

n	1	2	3	4	5	6	7	8
B_n	2	4	7	11	16	22	29	37

Evidently there is a connection with

$$B_n - 1 = \frac{1}{2}n(n + 1)$$

which is the sum of the first n numbers, as can be seen by writing out that sum both

forwards as $1 + 2 + 3 + \ldots + (n - 1) + n = S_n$ (say)
and backwards $n + (n - 1) + \ldots + 2 + 1 = S_n$

and adding which gives

$$S_n = \frac{1}{2}n(n + 1). \text{ Thus } B_n = S_n + 1.$$

72 Surprises

We first develop a 'surprise', in five steps (a)–(e), and then explain it. The steps, with four examples of each, are as follows.

(a) Write down a 3-digit number (in the 'hundreds') in which the difference between the first and last digits is 2 or more. The four different examples are

$$124\ 782\ 539\ 891.$$

(b) Reverse the digits in each, to give

$$421\ 287\ 935\ 198.$$

(c) Subtract the smaller from the larger in (a) and (b) to give

$$297\ 495\ 396\ 693.$$

Notice that these differences are all multiples of 99, namely

$$3 \times 99 \quad 5 \times 99 \quad 4 \times 99 \quad 7 \times 99.$$

(d) Reverse the digits in (c), to give

$$792\ 594\ 693\ 396.$$

(e) Add each pair of corresponding numbers in (c) and (d). This gives

$$1089\ 1089\ 1089\ 1089.$$

The 'surprise' is, of course, that we get the same result in each example.

How can we explain this?

Already at stage (c) the differences are all multiples of 99. To explain *that* we first observe that our familiar system of writing 3-digit numbers has the meaning that if A, B and C are any single digits, the product

$$ABC = A \times 100 + B \times 10 + C \times 1.$$

If we reverse these digits we get the number

$$CBA = C \times 100 + B \times 10 + A \times 1.$$

Subtracting any such pair of numbers gives

$$ABC - CBA = (A - C) \times 100 + (B - B) \times 10 + (C - A) \times 1 = (A - C) \times 100 - (A - C) \times 1 = (A - C) \times 99$$

so we always get a multiple of 99 at this stage.

We have to prove that if we add any 3-digit multiple of 99 to its reverse, we get 1089.

The complete list of 3-digit multiples of 99 is

$$1 \times 99 = 099, 2 \times 99 = 198, 3 \times 99 = 297, 4 \times 99 = 396, 5 \times 99 = 495, 6 \times 99 = 594,$$
$$7 \times 99 = 693, 8 \times 99 = 792, 9 \times 99 = 891, 10 \times 99 = 990.$$

Now we want to *prove* that if we add any 3-digit multiple of 99 to its reverse, we get 1089.

Observation of the list of 3-digit multiples above shows that they all have the same structure, namely that for any number N in the list N = 1,2,3,4,5,6,7,8,9,10, we find that

$$N \times 99 = [N - 1]9[10 - N].$$

(An example is $8 \times 99 = 792 = [8 - 1]9[10 - 8]$.)

Here $[N - 1]$ is in the hundreds column, 9 is in the tens column and $[10 - N]$ is in the units column, so the abbreviation means

$$[N - 1]9[10 - N] = [N - 1] \times 100 + 9 \times 10 + [10 - N] \times 1.$$

Reversing the digits gives $[10 - N]9[N - 1]$.
Adding $[N - 1]9[10 - N]$ to $[10 - N]9[N - 1]$
by the usual rules for vertical addition gives
$10 - N + N - 1 = 9$ in the units column, $9 + 9 = 18$ so 8 in the tens column, and 'carrying' 1, we have $1 + N - 1 + 10 - N = 10$ in the hundreds column, so we obtain 1089 in all for *any* N. This completes the proof.

From the fact, referred to above, that any 3-digit multiple of 99 has the structure

$$[N - 1]9[10 - N] \text{ for any } N = 1, 2, 3, 4, 5, 6, 7, 8, 9, 10$$

we can see that the *sum* of its digits is $N - 1 + 9 + 10 - N = 18$. That is, the sum of the digits in any 3-digit multiple of 99 is 18.

It is also the case that 1089 is a multiple of 99, namely

$$1089 = 1100 - 11 = (100 - 1) \times 11 = 99 \times 11.$$

Referring back to (a) above, there are smaller surprises. If the difference between the first and last digits of the starting number is only 1, the final outcome of the same sequence of steps (b) to (e) is always 198.

And if the starting digits are the same, the final outcome is 0.

David Acheson uses the 'surprise' described here about *1089 and all that* to begin his own book about 'A journey into mathematics' (Oxford University Press, 2002).

73 Sewell's Spirals

I should not really choose this title, because it might sound presumptuous, but the alliteration is convenient as a reminder. After exploring this topic (which I had not seen elsewhere) with children in class, I subsequently published a paper about it called 'Straight Sequences in a Spiralling Grid of Numbers' in the journal called *Mathematics in School*, Volume 38, Number 2, pages 15 to 17, March 2009.

We begin with a square grid of rectangular boxes, and we enter any chosen integer s (standing for 'starter') in a box which we deem to be the centre of the grid. This s can be positive or negative or zero.

Then we enter $s+1$ to the right of s, $s+2$ above $s+1$, then $s+3$ to the left of $s+2$, and $s+4$ to the left of $s+3$, then $s+5$ below $s+4$, and $s+6$ below $s+5$, then $s+7$ to the right of $s+6$, and $s+8$ to the right of $s+7$, thus completing an anticlockwise spiral of numbers in the eight boxes around the centre box.

Now enter $s+9$ to the right of $s+8$, and construct the next circuit with $s+10$, $s+11$, $s+12$ upwards from $s+9$, then $s+13$, $s+14$, $s+15$, $s+16$ to the left of $s+12$, then $s+17$, $s+18$, $s+19$, $s+20$ downwards from $s+16$, then $s+21$, $s+22$, $s+23$, $s+24$ to the right of $s+20$. This circuit occupies 16 boxes as illustrated in the diagram.

Now enter $s+25$ to the right of $s+24$, then $s+26$ upwards and so on round another circuit, which occupies 24 boxes. Continue to construct circuits in this way. Evidently, if we deem the first circuit to be the one which immediately surrounds the starting box, the mth circuit will contain $8m$ boxes.

.	.		.		.		.		.		.
.	$s+16$	$s+15$	$s+14$	$s+13$	$s+12$	.
.	$s+17$	$s+4$	$s+3$	$s+2$	$s+11$	.
.	$s+18$	$s+5$	s	$s+1$	$s+10$	.
.	$s+19$	$s+6$	$s+7$	$s+8$	$s+9$	.
.	$s+20$	$s+21$	$s+22$	$s+23$	$s+24$	$s+25$
.	.	.	.	.	.	.

Having constructed several circuits in this way (say eight, which is enough to reveal significant patterns), with any s, we wish to deduce formulae which describe the numbers which occupy the four half-diagonals from the centre, and also those which occupy the four horizontal and vertical half-lines from the centre.

Let us denote the wth box in the south-west (SW) half-diagonal by $w = 0$ (the centre box),1,2,3,... outwards from the centre. Let x, y, z be similarly used as counting numbers to denote the distance, from the centre, of boxes in the north-west (NW), north-east (NE) and south-east (SE) half-diagonals, respectively.

The topic of immediate interest from the point of view of developing this material in a way which can be explained to ten-year-olds, and which has a clear conclusion, is the method by which we can derive the four algebraic formulae which will reveal the occupants of the four sets of boxes on the half-diagonals as functions of these distances w, x, y, z out from the centre. In one case, for the SE half-diagonal, we shall need a first-order difference equation.

Then we shall use a, b, c, d to denote the coordinates of the boxes along the half-lines in the north (N), east (E), south (S) and west(W) directions, respectively, from the centre (with $a = b = c = d = 0$ there). We shall find that a single second order difference equation is needed to describe the occupants of these four half-lines.

Examination of the SW half-diagonal reveals its entries to be

$$s + 2w(2w + 1) \text{ in the } w\text{th box from the centre.}$$

The NW half-diagonal has entries

$$s + (2x)^2 \text{ in the } x\text{th box from the centre.}$$

The NE half-diagonal has entries

$$s + (2y - 1)2y \text{ in the } y\text{th box from the centre.}$$

The SE half-diagonal has entries

$$s + 4z(z + 1) \text{ in the } z\text{th box from the centre.}$$

Is it a surprise, given the spiral method of construction, that the four half-diagonals are described by such simple, but different, formulae?

Next we look at the entries in the four paths out from the centre along the four principal compass directions. Using a, b, c, d as coordinates in the N, E, S, W directions respectively, we find that the entries are described by the formulae

$$s + 4a^2 - a, s + 4b^2 - 3b, s + 4c^2 + 3c, s + 4d^2 + d.$$

Further questions arise. For example, if s is a prime number, for how long does the sequence of values of the half-diagonal numbers continue to be prime? Some answers can be given, as follows.

74 Prime Diagonals

The spiral construction in the previous section raises some interesting questions.

For example, if we start with $s = 17$, we find after nine circuits that the sequence of numbers in the SW-NE diagonal is

$$359, 289, 227, 173, 127, 89, 59, 37, 23, 17, 19, 29, 47, 73, 107, 149, 199, 257, 323.$$

It turns out to be the case that there is a large central batch of numbers in this sequence which are *prime* numbers, namely from 227 down to 17 and up again to 257.

But outside that range we find that $289 = 17 \times 17$ is actually a square number, and $323 = 17 \times 19$ is also not prime.

In the section on 'Don't jump to conclusions' we saw that $n^2 + n + 17$ is prime only as far as $n \leq 15$. And in that on 'Euler's formula' we saw that $n^2 + n + 41$ is prime for all integers up to 39, but not for $n = 40$.

So we have another illustration here that in mathematics it is unwise to jump to conclusions without being willing to search for a subsequent proof. By all means make an

Figure 62: Garden ornament having cubic horizontal sections

initial guess, if that suggests itself, but then make the effort to provide a proof which confirms that guess (or, perhaps, disproves it). That is how mathematics grows, by conjecture followed by proof.

75 Cubic Cusp in the Classroom

The outdoor photograph shows a garden ornament made by Richard Sewell in 2001, using the equations below. It consists of straight steel rods, of one centimetre diameter, which connect two different cubic curves. The ends of the long rods lie on two parallel planes one metre apart. The two cubic curves are approximated by short straight rods of appropriate (different) lengths. I exhibited the structure to children in the classroom.

The two cubics, each in an x, t plane spanned by cartesian coordinates, belong to the same family

$$x = (y - 2)t - t^3 \text{ for the particular values } y = 0 \text{ and } y = 8.$$

The diagrams show these two cubics

$$x = -2t - t^3 \text{ and } x = 6t - t^3.$$

The second cubic has turning points at $t = \pm\sqrt{2}$ where $x = \pm 4\sqrt{2}$. As the value of y is allowed to change from successive constant values between 0 and 8, the gradient

$$\frac{dx}{dt} = y - 2 - 3t^2$$

at the origin $t = 0$ is negative for $y < 2$ and positive for $y > 2$, and zero at $y = 2$, for which the cubic is $x = -t^3$.

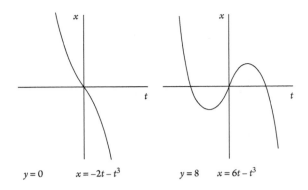

$$y = 0 \qquad x = -2t - t^3 \qquad\qquad y = 8 \qquad x = 6t - t^3$$

Figure 63: Cubic curves at each end of the garden ornament

When we think of t, x, y as cartesian coordinates spanning the three-dimensional space in which we live, the y-axis is the third coordinate, and

$$x = (y - 2)t - t^3$$

describes a surface. Each fixed value of t specifies a different straight line, and these lines show up as the long straight rods in the photograph. The cubic curves themselves (at $y = 0$ and $y = 8$) are approximated by two sequences of short straight rods joining the ends of the long rods.

The photograph shows a cusped shadow on the ground, which is the projected envelope of the straight rods. The equation of this cusped curve is

$$27x^2 = 4(y - 2)^3.$$

This is obtained by differentiating $x = (y - 2)t - t^3$ with respect to t, giving $(y - 2) = 3t^2$ and then eliminating t from these two equations.

A classroom demonstration can be achieved by plotting the two cubics on graph paper, sticking those graphs onto stiff cardboard sheets, piercing the cards in appropriate places to accommodate string or cotton threaded through between matching holes, wedging the two cards an appropriate distance apart according to the scale being used and then threading cotton through corresponding holes.

Further explicit details can be found in an article which I wrote called 'Mathematics in the Garden', published in *Mathematics Today* **42**, pages 215–216, December 2006.

Many attractive applications of the properties of the cubic curves mentioned here, and the associated geometry, have been pioneered by Zeeman (1977), Poston and Stewart (1978) and others.

76 Nature's Circles

These are numerous, and aesthetically satisfying. Among the most obvious are the outlines of the Sun and the full Moon, which are whole circles as seen from Earth.

When a stone is thrown into a still pond, the impact causes a disturbance which generates a sequence of outgoing circular waves. This illustrates how a localised disturbance in any homogeneous two-dimensional medium may propagate outwards as a circular wave.

During Autumn it is not uncommon to find what are called 'fairy rings' on grassy areas, such as garden lawns. These evidently begin from a central nucleus which sends out underground messages (via threads of so-called *mycelia*) in all directions, until they decide to manifest themselves above ground in a circle of fungi, or part of a circle, on the surface.

The photograph shows a semicircular fairy ring which grew in my own garden, fortuitously next to the cubic cusp model which is referred to in the previous section.

77 Rainbow

The photograph of a rainbow over Malmesbury in Wiltshire was taken by Robert Peel, and it appeared in *The Daily Telegraph* on 13 December 2011. It is reproduced here by agreement with SWNS in Bristol.

The image may, at first sight, seem to be part of a circle, but closer examination shows that it is not. We prove this as follows. Ignoring the secondary rainbow (with reversed order of colouring) above the main rainbow, we select the yellow band in the middle of the main rainbow. This band is not much thicker than would be an arc drawn by pencil. We mark three points, unsymmetrically, on this yellow band, and join them by two successive straight chords.

Figure 64: Cubic cusp and fairy ring

Figure 65: Rainbow and circular construction

The perpendicular bisectors of those two chords intersect in a point which will be the centre of a circle passing through the chosen three points on the yellow arc (each of these two bisectors is the join of the intersections of a pair of short circular arcs which are centred, as shown, on an adjacent pair of the chosen points).

The circular arc thus drawn can be seen to be superimposed on the yellow arc in the centre of the picture, but we also see that it departs from the rainbow (and comes below it) towards the two ends of the rainbow.

This geometrical construction proves that the photographic image of the whole rainbow is not circular. It uses the geometrical exercise which we introduced in Section 21. I described this construction in the Autumn 2012 maths newsletter of The Royal Institution of Great Britain.

However, this construction does not prove that the real rainbow itself is not circular. The photograph may have been taken with a 'wide-angle' lens, which can introduce distortion of the image towards the edges of the picture. The reader will be able to find more information about the formation of rainbows on widely available web-sites.

78 Basis and Bases of Arithmetic

This title is intended to encapsulate the fact that the basis of arithmetic can appear in various forms, depending on which 'base' we choose to work in.

'Arithmetic' is about the relations between numbers.

Our usual arithmetic, which we meet first as young children, and then in primary school, uses *ten* different symbols 0 1 2 3 4 5 6 7 8 9 to describe or write *all* numbers. It is called the *denary* system (after the Latin for 'ten') or 'base ten'. The layout is to imagine rows of slots, and then to attach a value to each slot as in the following table for the examples 10, 324 and 59 867. Each dot entered in this and following tables represents a blank space.

ten thousands	thousands	hundreds	tens	units
.	.	.	1	0
.	.	3	2	4
5	9	8	6	7

We could use *powers* to rewrite the column heads as

$$10^4 = 10\,000, 10^3 = 1000, 10^2 = 100, 10^1 = 10, 10^0 = 1.$$

Thus, for example, in base 10, the meaning of numbers is illustrated by

$$324 = 3 \times 10^2 + 2 \times 10^1 + 4 \times 10^0.$$

Addition by columns uses *carrying* from one column to the next.

But the familiar or 'usual' base ten arithmetic is not the only possible type. *Computer* arithmetic uses only *two* different symbols, 0 and 1, to write *all* numbers. This is called the *binary* system (from the Latin for two) or *base two* arithmetic.

It is useful in computing because the presence of only *two* options can represent *on/off* or *yes/no* repeatedly.

The layout is again to imagine *rows of slots*, and then attach a value to each slot, but differently from the denary system. An example is 1,2,3,4,5,6,7,8 written is binary as

eights $8 = 2^3$	fours $4 = 2^2$	twos $2 = 2^1$	units $1 = 2^0$
.	.	.	1
.	.	1	0
.	.	1	1
.	1	0	0
.	1	0	1
.	1	1	0
.	1	1	1
1	0	0	0

As an illustration, the following is a simple vertical addition sum, using both binary first,

2^3	2^2	2^1	2^0
.	.	1	1
.	1	0	1
1	0	0	0

and denary second.

10^2	10^1	10^0
.	.	3
.	.	5
.	.	8

In the classroom children can be asked to exhibit various (small) numbers in binary on their (eight) fingers (leaving out thumbs), by holding a finger up to signify 1, and down to signify 0. Thus using both sets of four fingers simultaneously, in order, we can signal

$$2^0 = 1, 2^1 = 2, 2^2 = 4, 2^3 = 8, 2^4 = 16, 2^5 = 32, 2^6 = 64, 2^7 = 128.$$

Other bases beside base 10 (denary) and base 2 (binary) can be referred to. For example, base 60 has been associated with the Babylonians over 2000 years ago. This persists in our time-measuring system today, because we have $60^0 = 1$ second, $60^1 = 60$ seconds = 1 minute and $60^2 = 3600$ seconds = 60 minutes = 1 hour.

This base sixty system also persists in angle measurements, because we define a full circle to have $360 = 6 \times 60$ degrees. Subdivisions include the terminology that 1 degree = 60 minutes of arc, and 1 minute = 60 seconds of arc. These words do not have the same meaning as in the time measures, of course.

79 Lunes

This section discusses a problem which has two parts. We shall see that the time which elapsed between the solution of the two parts was about 2420 years.

The hypotenuse (of length c, say) of any right-angled triangle is a diameter of a circle which circumscribes that triangle, because the angle in a semi-circle is a right angle. Two more circles can be drawn using the shorter sides of the triangle as diameters, with lengths a and b (say). This construction creates two *lunes* outside the starting circle. The boundary of each lune is a pair of circular arcs. In each pair, one arc is a semi-circle and the other is smaller than a semi-circle. Thus each *external* lune (external to the starting circle) looks like a crescent Moon ('lunar' means a property of the Moon, here its shape).

In the diagrams, by Jacqueline Fairbairn, six different regions are of particular interest. These regions are differently coloured.

The areas of the coloured regions are designated by the letters O (brown), P (purple), Q (yellow), R (blue), S (green) and T (red), as follows.

The starting circle has area O. The right-angled triangle has area T. The external lunes have areas R and S. The construction also creates a pair of internal lunes (inside the starting circle) having areas P and Q, and they overlap each other (unlike the external lunes).

The area of the triangle is $\frac{1}{2}ab = T$. The areas of the two external lunes are $\frac{1}{2}\pi\left(\frac{a}{2}\right)^2 - \alpha = R$, and $\frac{1}{2}\pi\left(\frac{b}{2}\right)^2 - \beta = S$, where α and β are the areas of those interior parts of the

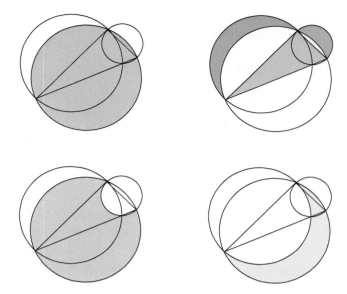

Figure 66: Areas of lunes

starting circle bounded by sides a and b of the triangle, respectively. Hence $R + S = \frac{\pi}{8}(a^2 + b^2) - (\alpha + \beta)$. But $\alpha + \beta = \frac{\pi c^2}{8} - \frac{1}{2}ab$, and therefore, by Pythagoras' Theorem, $R + S = \frac{1}{2}ab = T$. This proves Theorem 1:

Theorem 1: *The sum of the areas of the two exterior lunes is equal to the area of the right-angled triangle.*

This theorem is attributed by Pickover, in an attractive book (2009), to Hippocrates (c. 470 B.C.–c. 400 B.C.) of Chios. So it is at least possible that Hippocrates knew the theorem of Pythagoras (c. 570 B.C.–c. 495 B.C.) of Samos which is used here. Chios and Samos are islands in the Aegean Sea only 50 miles apart (Pythagoras also worked in southern Italy). Pickover leaves the topic at this point, and moves onto other things in his book.

However, we can also observe that the construction of the two circles on the shorter sides of the triangle *also* has the effect of creating the other two (and larger) lunes, having areas P and Q, which are interior to the starting circle, and which overlap each other. A new question arises, which I have not seen posed before. What can we say about the areas of these *internal* lunes? The answer to this is as follows.

Theorem 2: *The sum of the areas of the two internal (and overlapping) lunes is equal to the whole area within the starting circle plus the area of the right-angled triangle based on its diameter. That is, P + Q = T + O. Areas of overlapping regions are counted twice.*

The proof is as follows. Since P and Q are the areas of these internal lunes, and with the previous α and β, these areas satisfy

$$Q + \alpha + \pi \frac{a^2}{8} = \pi \frac{c^2}{4} = P + \beta + \pi \frac{b^2}{8}.$$

Adding these two equations gives $P + Q = \pi \left(\frac{c}{2}\right)^2 + \frac{1}{2}ab$ as required.

My research (Ph.D.) supervisor at Nottingham University over 50 years ago (1957–1960), the late Professor Rodney Hill F.R.S., once told me that one should be thinking about one's research problem 'all the time, even in the bath'. I was reminded of this when, with little further preparation, the following extrapolation of the foregoing lune geometry came to me in the middle of the night. [My biography of Hill has been published by The Royal Society (2015).]

Theorem 3: (The 3.10 a.m. Theorem) *The area of the starting circle is the difference between the combined areas of the two internal lunes and the combined areas of the two external lunes. That is,* $O = (P + Q) - (R + S)$.

We notice that the area of the right-angled triangle which began this investigation is not mentioned explicitly in the statement of Theorem 3.

A version of the material in this section was published in *The Mathematical Gazette* in March 2014, pages 129–131.

I used the coloured Lunes diagram as my 2013 Christmas card, with the explanations that:

Red area = Blue area + Green area (Hippocrates of Chios, 410 B.C.);
Purple area + Yellow area = Red area + Brown area (Michael Sewell, 2013 A.D.).
So 2423 years separate the two results.

80 An Octet of Equal Circles

In this section we describe some properties of a sequence of circles which all have the same diameter.

Suppose two such circles intersect. There are three possible cases. Each centre may lie within the other circle, or each centre may lie on the circumference of the other circle, or each centre may lie outside the other circle. In the last case the two circles either touch at one common point, or intersect in two points.

For brevity here we develop some properties of the very last case when a third circle is superimposed on the first two, so that it intersects both.

There can be six possible distinct locations for the centre of the third circle: (a) within both of the first two circles; (b) within one, and on the circumference of the other; (c) within one and outside the other; (d) outside one and on the other; (e) on an intersection of both; (f) outside both.

The locations (b), (d) and (e), in which the third centre is *on* the circumference of one or both the first two circles, can be regarded as non-typical in an obvious sense. Therefore we leave those three aside, and regard the three locations (a), (c) and (f) as more typical.

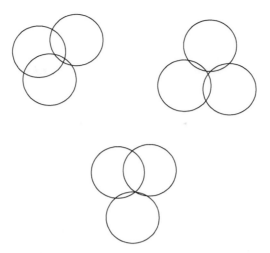

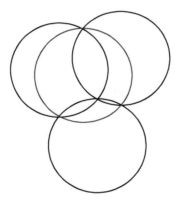

Figure 67: Three intersection cases (g) and (i) [above] and (h) [below]

Figure 68: Johnson's circle

For the sake of brevity we pursue here the consequences further for just *one* of those cases, and we choose the case (f). The reader is invited to explore the other cases.

The case (f) has two possibilities for the location of the third circle. Its centre might be on the chord ('produced' outside the circles) which joins the intersections of the first two circles (the symmetric case), or off it (the non-symmetric case).

Now we investigate further the non-symmetric case just described. For this case there are three ways in which the third circle may intersect the first two (and each twice), as follows. Either (g) each of one intersection is *within* the other circle; or (h) each is *at* an intersection of the first two, so that the three circles have one point in common; or (i) both intersections

Figure 69: Four pairs of congruent triangles

are *outside* the zone common to the intersection of the first two circles. We show these last intersecting three cases in the first trio of diagrams.

In a theorem of 1916 by R.A. Johnson, quoted by Pickover (2009, page 332), a *further* property of case (h) is that when three identical (but not necessarily symmetrically disposed) circles pass through a common point, their other three pairwise intersections must lie on *another* (fourth) circle which has the same size as the first three. This fourth circle is also shown in the next diagram. (The reader is recommended to repeat such a diagram). I noticed when I drew this diagram that there are more symmetries, which I had not seen described elsewhere. They are as follows, and they lead to the octet which is the substance of this section.

First we notice that *any* three of the four circles now displayed could be regarded as the first three circles having a common intersection point as in the discussion, and the fourth circle passes through the three pairwise intersections of those chosen first three. Thus there are really *four* examples of Johnson's Theorem in the same diagram. In other words, if the centres of the starting three are labelled A, B, C and the fourth one has centre J, we could equally well have had A, B, J as the starting trio and then C as the fourth one; or A, C, J as the first trio and B as the fourth; or B, C, J as the first trio and A as the fourth.

This leads us to observe more properties of such a diagram. If we draw on the diagram the triangle whose vertices are at the centres of the *first* three circles which have the common intersection required of them, that triangle is neither equilateral nor isosceles in the general case. Even so, we can see from the construction that that this triangle is congruent to a *second*

Figure 70: Octet of circles

triangle which can be drawn, whose vertices are at the three other and pairwise intersections of the first three circles. In addition, this construction shows that the corresponding pairs of sides of those two triangles are parallel. Also, the joins of centres of each pair of circles are equal in length to the chord of the third circle which meets this pair. Thus we finally have four different pairs of congruent triangles, as illustrated.

This means, by symmetry, that the circle which circumscribes the second triangle is the same size as the first trio of circles and as the circle which passes through their centres. This further means, by symmetry, that a *second* quartet of circles can be constructed, having the same size as those of the first quartet, but reversed in orientation. One of them circumscribes the first triangle, and the other three have their centres at the vertices of the new triangle. An additional circle of the same size circumscribes the second triangle.

Thus we have finally not merely four identical circles, initially intersecting as described in Johnson's Theorem, but *eight* circles having such identical properties. These are all shown in the last diagram. This is the 'Octet' named in the title of this section. Different colours help clarity.

I have described these results for the octet in an article published in *The Mathematical Gazette* in November 2015. Johnson's original (1916) proof for his quartet is as follows. In this statement of his proof, I have replaced Johnson's lettering C_1, C_2, C_3 by A, B, C respectively; and his P_1, P_2, P_3 by P, Q, R respectively. Each circle can be regarded as labelled by its centre.

'Denote the centers of the circles by A, B, C, the intersections of B and C by O and P, those of C and A by O and Q, those of A and B by O and R. Then OBPC is a rhombus, and so is OCQA. Hence, BP and AQ are equal, ABPQ is a parallelogram and PQ is equal to AB. Thus the triangles ABC and PQR are congruent, and have equal circumcircles. But the circumcircle of the former has its center at O, and is equal to each of the given circles. Hence, the circle through PQR is equal to each of the given circles.'

The centre J of the Johnson circle for A,B,C is not mentioned in his proof of the theorem. (The American spelling of centre is center.)

81 Alternative Construction of the Octet

There is an alternative construction of the octet (and with different suggested colouring), as follows, which the reader may find informative to emulate. First draw the three equal circles (coloured blue, say), centred at A, B, C, but not symmetrically disposed, and having a three-way intersection at O. Construct their Johnson circle (say green) centred at J such that it passes through their three pairwise intersections. Use O to be the centre of a fifth circle (say red) of the same size.

Make a transparency of the quintet constructed thus far. Rotate the first diagram of the quintet by 180 degrees, and lay it upon the original quintet in such a way that the first O is directly upon the second J, and the first J is directly upon the second O. The green and red circles will then be found to coincide, thus providing two circles in the centre of the diagram; and outside them will be six blue circles. The resulting composite diagram provides the octet which is the subject of the previous section.

We could say that there is an intrinsic *duality* present in the octet diagram, in the following sense. The first trio of circles centred at A, B, C, which have a common intersection at O, and whose Johnson circle has centre J, generates the second trio of circles centred at P, Q, R, which have a common intersection at J, and whose Johnson circle has centre at O.

82 Triangle Constructions

The reader may care to confirm some triangle properties indicated in the first of the two sections immediately above, by carrying out the following constructions.

Confirm by drawing the diagram that the triangle ABC which joins the centres of the starting trio of circles is congruent to the triangle PQR which joins their pairwise intersection points. Check that the two triangles are reversed in orientation, with pairs of corresponding sides parallel, i.e. AB = PQ and are parallel, BC = QR and are parallel, and CA = RP and are parallel.

(a) Construct a diagram which shows the pair of triangles ABC and PQR. Verify by measuring the diagram that pairs of corresponding sides are in fact parallel, so that the two triangles are congruent and are, in three ways, pairwise reflections of each other.

(b) Repeat the construction for the pair of triangles ACJ and PRO.

(c) Repeat the construction for the pair of triangles BCJ and QRO.

(d) Repeat the construction for the pair of triangles ABJ and PQO.

The foregoing constructions are focussed on the four circles with centres A, B, C and J (the last one being the Johnson circle through the pairwise intersections of the first three).

83 A Mosaic of Equal Circles

This section is offered as an exercise for the reader, to make some hopefully attractive geometrical constructions. They provide patterns which, as far as I am aware, are quite novel. The Johnson circle described above added a fourth equal circle to a trio of ones already there which pass through a single point. The octet which I discovered in 2015 was obtained by noticing a kind of mirror symmetry which is an added consequence of the Johnson quartet.

Having drawn the octet, it can be seen that this initial network of eight equal circles can be extended indefinitely over the whole plane, thus providing what can be called a *mosaic* of equal circles. This is achieved as follows.

The circles centred at A, R, B, P, C, Q have six pairwise *outer* intersections; those points can be used as the centres of six more equal circles of the same size, which pass through AR, RB, BP, PC, CQ and QA. Then *their* outer intersections can be used to be the centres of six more equal circles of the same size, making twenty circles thus far (if we were to use the inner intersections as well to construct more circles, this would make the diagram more complicated than might be legible). This process can be continued by using the *new* set of outer intersections, which will have been created, as centres for more circles, and so on.

Thus far the development of the number of circles in the sequence stated explicitly above can be summarised by the following two series. The second series lists the centres used to achieve the numbers of circles in the series above it.

$$3 + 1 + 1 + 3 + 6 + 6 + 12 + \ldots$$
$$ABC + J + O + PQR + (AR + AB + BP + PC + CQ + QA) + \ldots + \ldots + \ldots$$

The reader is invited to construct the diagram showing the implied multiple intersections. It would be possible to extend the diagram beyond the first twenty, but there arises the danger of running off the page.

By now it is clear that the whole plane can be covered by an *infinite* mosaic of circles constructed in the stated way.

84 Intersection of Equal Spheres

If we move from two dimensions into three, to enquire if the ideas in the foregoing four sections can be generalised in that direction, we shall be dealing with sets of equal spheres instead of with sets of equal circles. This enters territory which has never been explored, as far as I know.

The reader is invited to think about it. For example, two equal spheres can intersect in a circle (generalising the fact that two equal circles in one plane can intersect in two points when suitably positioned). Where should a third equal sphere be placed so that it intersects the first two in a way which is *generic* in a way to be defined (rather than as some special case)? Can more spheres be added in those three dimensions to create a generalisation, for the spheres, which corresponds to the Johnson circle in the plane diagram, and to generalisations of it which were indicated above (e.g. of the octet, or of the mosaic of circles)?

85 Christmas Cracker

As an entertainment just before Christmas, a question and answer session with ten children eventually produced the following table.

French	German	Italian	Number
un	eins	uno	1
deux	zwei	due	2
trois	drei	tre	3
quatre	vier	quattro	4
cinq	fünf	cinque	5
six	sechs	sei	6
sept	sieben	sette	7
huit	acht	otto	8
neuf	neun	nove	9
dix	zehn	dieci	10

We can use this information first to construct some addition and subtraction sums, such as

$$\text{dix} + \text{neun} - \text{dieci} = 9 \text{ and quatre} + \text{eins} - \text{tre} = 2;$$

and then multiplication and division sums such as

$$(\text{cinq} \times \text{acht}) + \text{otto} = (5 \times 8) + 8 = 48, \text{and}$$
$$\text{dix} - (\text{acht} \times \text{due}) = 10 - (8 \times 2) = -6.$$

86 Ostrich Egg

The equation of a circle having radius a is $x^2 + y^2 = a^2$, using cartesian coordinates x and y, or $(x/a)^2 + (y/a)^2 = 1$. An ellipse is like an elongated circle, and in the simplest form the equation of an ellipse is $(x/a)^2 + (y/b)^2 = 1$, where a and b are the lengths of the half-axes. If $a > b$, the x-axis is called the major axis and the y-axis is called the minor axis. The planets move around the Sun in orbits which are ellipses, aside from small secondary effects.

The generalisation of an ellipse to three dimensions is a surface called an ellipsoid. In terms of cartesian spatial coordinates x, y, z the simplest (so-called canonical) equation of an ellipsoid is $(x/a)^2 + (y/b)^2 + (z/c)^2 = 1$, where a, b, c are the given (constant) lengths of the so-called semi-axes of the ellipsoid.

There are particular cases which have $a = b < c$, called a prolate ellipsoid (thinner in two equal directions than the third); and also $a = b > c$, called an oblate ellipsoid (thicker in two equal directions than the third).

Birds eggs usually have rotational symmetry about their longest axis (presumably the most convenient direction in which the egg was actually laid). It may be rather infrequent that eggs emerge with a shape precisely that of a prolate ellipsoid (because there is sometimes discernable asymmetry about the smaller and circular cross-section).

But I did buy an ostrich egg at a farm shop, which is 12 cm wide and 16 cm long, and which does seem to have the symmetries which make it a close approximation to a prolate ellipsoid. The minor and major semi-axes are therefore $a = b = 6$ cm, and $c = 8$ cm, respectively. The photograph shows it, supported in an egg cup designed for a hen's egg, which emphasises the relative sizes.

87 Holditch's Theorem

This applies to any smooth closed convex curve, which we call C. A straight chord of fixed length is drawn anywhere inside the curve. Imagine that this chord is then slid around inside the curve so that its two ends always touch the curve. Choose any fixed point P on the chord, which divides the chord into two fixed lengths p and q. As the curve is moved right round C, this point P on it will trace another closed curve (say Q) within C. Holditch's Theorem, dating from 1858, states that the area between the curves C and Q is πpq, whatever be the shape of C.

This remarkable result is quoted, without proof, by Pickover (2009, p 250). Hamnet Holditch (1800–1867) was President of Gonville and Caius College in Cambridge.

The proof for the special case when C is a circle, of diameter $2R$ say, is relatively simple, as follows. A chord of fixed length $2p < 2R$ is chosen. This chord is now slid around the circle. It therefore acts as a tangent to, and so envelops, a smaller circle c (say) which has the same centre as C. Thus this chord is tangent to c in every position. The radius of c is the square root of $R^2 - p^2$ (say r), by Pythagoras' Theorem, because p is the half-length of the sliding chord.

The area of the smaller circle is $\pi r^2 = \pi(R^2 - p^2)$, so the area of the circular band between the two circles is $\pi R^2 - \pi r^2 = \pi p^2$, which is the same as the area of a circle of radius p. This quantity πp^2 is what πpq in the first paragraph, for the more general case, reduces to in this special case.

Figure 71: Ostrich egg

88 A Coffee Shop Problem

It is explained in the Introduction that this book evolved during a period of ten years during which I devised and delivered Masterclasses in a primary school. The material is all original in the sense of being constructed specifically for that purpose, and not copied from elsewhere. The style used in class was a question and answer technique, and not via formal 'exercises' or homework. That is why such additional problems do not appear in the text, because they were not used. The book aims, like the lessons themselves, to bring out important mathematical ideas in a lucid and enjoyable way which encourages reader participation in each section.

In this section I will present an investigation which, after initial illustrations, leave the reader with something to do which hopefully will be somewhat testing, but agreeable. Enjoyment of mathematics is a worthwhile objective, and one which is a common experience.

This 'coffee shop problem' emerged as exactly that. It was posed to me in a coffee shop by my wife, who is an experienced teacher of mathematics, currently in a primary school, and previously in a secondary school.

The problem as originally posed is as follows. Eight equally spaced points around a circle can be viewed as located at N, E, S, W and NE, SE, SW and NW compass positions on the circumference. How many *different* triangles can be drawn whose three vertices (corners) are at three of those points? Reflected triangles of the same shape are *not* to be regarded as different, nor are rotated triangles of the same shape.

I was able to solve this problem, under supervision, without requiring a second cup of coffee. The answer is five. The reader is invited to draw all the diagrams, as an exercise.

But then a mathematician is trained to always stop and think again. Have I really finished? Can the problem be generalised?

It can, but perhaps not on the basis of only *one* cup of coffee.

A generalisation can proceed as follows. Let p points be equally distributed around a circle, at equal intervals of $360/p$ degrees. How many (say n) distinct polygons having q ($< p$) sides can be drawn which have vertices at q (= 3 or > 3) of those p points? Again reflections and rotations are not to be counted as distinct cases.

Plainly, when $q = p$ the answer is $n = 1$, with a regular polygon of p sides. For example, $q = p = 3$ provides an equilateral triangle, and $q = p = 4$ provides a square. For $q = 3$ with $p = 4$ the solution is an isosceles right-angled triangle.

Those cases each have a unique solution. Important questions in more advanced mathematics are whether a solution to a problem exists (if not it would be futile to look for it); and whether a solution, when known to exist, is unique. For example, the compression of a straight strut under an axial load, like a walking stick, is not unique if one presses too hard, because it will then buckle sideways, so providing a second solution to the problem of finding what deformation occurs under the axial load, in addition to the straight compression.

Less easy issues now arises in our problem in cases having $q < p$ and $p > 4$. The reader is invited to solve these for the cases in which, in turn, are $p = 5, 6, 7, 8$. (Sewell and Sewell, 2016). It will help considerably to draw the diagrams for each case; this will provide both confidence and confirmation.

An associated worthwhile question in each case is to check how many (say s) of the polygons are symmetric about at least one diameter of the circle.

This is not a short exercise, and it may be best to carry it out in stages, rather than all at one sitting.

The table of results is as follows.

p	5	5	6	6	6	7	7	7	7	8	8	8	8	8
q	4	3	5	4	3	6	5	4	3	7	6	5	4	3
n	1	2	1	3	3	1	3	4	4	1	4	4	8	5
s	1	2	1	3	2	1	3	3	3	1	4	2	6	3

89 Step Waves

When we keep our eyes open there are numerous interesting patterns in nature to be noticed. This book has described some of them, and provided signposts to the associated mathematics which underlies many of them. For a final example the photograph shows a 'step wave' pattern which I noticed on my sloping home street after heavy rainfall. The water is running, from top to bottom in the photograph, down the street which has a gradient of about 1 in 10 (so mildly steep). We can see that the stream forms itself into a sequence of stable steps, of which the picture shows 20, each having a height of about 1 cm, and a length of about 20 cm between each pair of steps, towards the observer at the bottom of the photograph. After the rainfall has ceased, a periodic pattern of debris which was carried by the stream remains for the alert observer to see. A mathematical explanation has been provided by Merkin and Needham (1984, 1986), as referenced below. It is more sophisticated than is appropriate for inclusion in this book, but the fascination which nature offers in such patterns commends itself to enquiring minds, as we have attempted to illustrate in the foregoing pages.

References

Acheson, D. *1089 and All That*. Oxford University Press 2002.
Brannan, D.A., Esplen, M.F. and Gray, J.J. *Geometry*. Cambridge University Press 1999.
Bryant, J. and Sangwin, C. *How Round Is Your Circle?* Princeton University Press 2008.
Durell, C.V. *Modern Geometry*. Macmillan 1920.

Figure 72: Step waves on a sloping street

Figure 73: Periodic debris pattern

Hardy, G.H. *A Mathematician's Apology*. Cambridge University Press, 1940.

Merkin, J.H. and Needham, D.J. On Roll Waves Down an Open Inclined Channel. Proc. Roy. Soc. Lond. A**394**, 259–278, 1984.

Merkin, J.H. and Needham, D.J. An Infinite Period Bifurcation Arising in Roll Waves Down an Inclined Channel. Proc. Roy. Soc. Lond. A**405**, 103–116, 1986.

Pickover, C.A. *The Math Book*. Sterling, New York 2009.

Poston, T. and Stewart, I. *Catastrophe Theory and its Applications*. Pitman 1978.

Preater, J. Letter on The Rugby Riddle. Mathematics Today **41**, 163, 2005.

Sewell, M.J. *Mathematics Masterclasses: Stretching the Imagination*. Oxford University Press 1997.

Sewell, M.J. A Rugby Riddle. Mathematics Today **41**, 126–128, 2005.

Sewell, M.J. The Rugby Riddle Revisited. Mathematics Today **41**, 196, 2005.

Sewell, M.J. Mathematics in the Garden. Mathematics Today **42**, 215–216, 2006.

Sewell, M.J. Straight Sequences on a Spiralling Grid of Numbers. Mathematics in School, **38**, No. 2, 15–17, 2009.

Sewell, M.J. Pythagoras and the Fleet Air Arm. Royal Institution of Great Britain Mathematics Masterclass Newsletter, Autumn 2010.

Sewell, M.J. Leavers from an Expanding School. Mathematics Today **49**, 137, 2013.

Sewell, M.J. Areas of Lunes. The Mathematical Gazette. **98**, No. 541, 129–131, March 2014.

Sewell, M.J. An Octet of Circles. The Mathematical Gazette. **99**, 468–473, November 2015.

Sewell, M.J. Rodney Hill. Biogr. Mems. Fell. R. Soc. **61**, 161–181, 2015.
Sewell, Michael and Sewell, Bridgid. A Coffee Shop Problem. Reading University Mathematics Department Report. February 2016.
Wells, D. *The Penguin Dictionary of Curious and Interesting Geometry*. Penguin 1991.
Young, L. *Mathematicians and Their Times*. North-Holland Publishing Company 1981.
Zeeman, E.C. *Catastrophe Theory: Selected Papers 1972–1977*. Addison-Wesley 1977.

Author information

Michael Sewell is Emeritus Professor of Applied Mathematics (since 1999) at the University of Reading. He was educated at The King's School in Grantham (1943–53: Head Boy and Captain of Cricket), where the influence of Sir Isaac Newton as a former pupil 300 years before was still emphasised. He entered the University of Nottingham (Mining Engineering 1953–54, and Mathematics in which he graduated B.Sc. (First Class Honours 1957), Ph.D. (1960, supervised by Professor Rodney Hill, F.R.S.) and D.Sc. (1974)).

He taught at the Universities of Nottingham (Mathematics 1960–1963), Bristol (Theoretical Mechanics 1963–1966), Reading (Reader in Applied Mathematics 1966–1978, Personal Professor 1978–1999) and Surrey (Engineering and Physical Sciences 2010).

He held Visiting Professorships at The University of Wisconsin–Madison 1970–71 and for shorter periods in 1982–84; Ecole Polytechnique de Paris 1976, University of Waterloo

Figure 74: Michael Sewell

Figure 75: Michael Sewell by Richard Sewell:3D printing

- Ontario 1978, the University of Mexico 1978 and the Cambridge University Newton Institute in 1996.

He has written more than 100 articles on topics in the mechanics of solids and fluids, duality, catastrophe and singularity theory, meteorology, mathematical education, family history and other areas. These include four published books:

Mechanics of Solids: The Rodney Hill 60th Anniversary Volume (edited with H.G. Hopkins), Pergamon Press 1982, 693 pp. (22 contributors).

Maximum and Minimum Principles; a unified approach, with applications. Cambridge University Press 1987, reprinted 1990 and 2007, 468 pp.

Mathematics Masterclasses - Stretching the Imagination (edited, with 13 contributors), Oxford University Press 1997, 233 pp.

Second Innings: RUASCC Passes Fifty (with Andy Eagle), 2012, 242 pp.

Professor Sewell was the Local Organiser of The Royal Institution Mathematics Masterclasses in Berkshire from 1990 to 1999, an annual initiative at the Department of Mathematics and Statistics of the University of Reading. He organised 114 3-hour Masterclasses (for 35 pupils each session) by invited visiting speakers during those years. The 1997 book above contains a selection of them. This Masterclass activity is still continuing, currently at Holyport College.

He played for Reading University Academic Staff Cricket Club from 1967 to 2002 (often opening the batting, and he was Captain 1978–1981). The 2012 book mentioned above is a history of the seasons from 1983 to 2012.

He is married to Bridgid (née Mowll), and has three sons, a grand-daughter and three grandsons. The plastic three-dimensional image (Figure 75) was made on the basis of a series of 30 photographs taken from a succession of different directions at intervals of 5 degrees.

INDEX